Arguments that Count

Inside Technology

edited by Wiebe E. Bijker, W. Bernard Carlson, and Trevor Pinch

Arguments that Count

Physics, Computing, and Missile Defense, 1949–2012

Rebecca Slayton

The MIT Press
Cambridge, Massachusetts
London, England

This book was set in Stone Sans and Stone Serif by the MIT Press.

Library of Congress Cataloging-in-Publication Data

Slayton, Rebecca, 1974-
Arguments that count : physics, computing, and missile defense, 1949-2012 / Rebecca Slayton.
    pages cm
Includes bibliographical references and index.
ISBN 978-0-262-01944-6 (hardcover : alk. paper)
ISBN 978-0-262-54957-8 (paperback)
1. National security—United States—History—20th century. 2. National security—United States—History—21st century. 3. Ballistic missile defenses—United States—History. 4. Physicists—Political activity—United States—History. 5. Computer scientists—Political activity—United States—History. 6. Physics—Political aspects—United States—History. 7. Computer science—Political aspects—United States—History. 8. Technological complexity—Political aspects—United States—History. 9. Software engineering—Political aspects—United States—History. 10. United States—Military policy. I. Title.
UA22.S57   2013
358.1'740973—dc23
2012051748

Contents

Acknowledgments

It is a privilege and a joy to thank the many people who have made this book possible. I began this project with the guidance and support of David Kaiser and Hugh Gusterson, who hosted my National Science Foundation postdoctoral project in MIT's Science, Technology, and Society (STS) Program. I am grateful for David's and Hugh's mentorship not only in the initial phases of my research, but also in the decade that followed. I am also thankful for the many scholars who sharpened my work during my time in MIT's STS program, including Anita Chan, Joe Dumit, Eden Medina, David Mindell, Rachel Prentice, Jenny Smith, Peter Shulman, Susan Silbey, and Roz Williams.

I turned my postdoctoral project into a book while working first as a Science Fellow and then an Affiliate of Stanford's Center for International Security and Cooperation (CISAC). I am indebted to many scholars at CISAC who helped me expand and clarify my research. First and foremost, Lynn Eden has proven to be one of my fiercest critics and staunchest supporters, reading numerous drafts, and always finding time to meet despite her packed-full schedule. Lynn taught me how to write a book, and I am deeply grateful. David Holloway has also given generously of his time, reading and offering insightful comments on several portions of the book manuscript. Barton Bernstein pointed me to the archives time and again, and combed meticulously through chapter drafts and redrafts. Charles Perrow, Stanford's "snowbird," has inspired my analysis, read parts of the book, and offered helpful comments for revision. I am also thankful for the support and insights of John Downer, Anne Harrington, Alex Montgomery, Pavel Podvig, Brenna Powell, Karthika Sasikumar, and Jan Stupl. Members of Stanford's Science, Technology, and Society program, including Robert McGinn, Eric Roberts, Selma Sabanovic, and Fred Turner also offered useful conversations and encouragement along the way.

I finished writing and revising this book in the History of Science and Technology Program at the University of Minnesota, and I am grateful for the encouragement of many people there. Tom Misa read several chapters and offered comments that significantly improved the argument and writing. I am also grateful for the collegial support of Jennifer Alexander, Bonnie Gidzak, and Sally Kohlstedt. And I thank Deb Dickey, Manny Esguerra, Tom Haakensen, Melanie Homan, and Julianne Kaul for helping make Minneapolis home for a year.

The research in this book would have been impossible without many people who graciously offered their time in interviews and e-mail conversations. Within the physics community, Richard Garwin, Marvin ("Murph") Goldberger, Kurt Gottfried, Spurgeon Keeney, George Lewis, Brockway McMillan, Ted Postol, George Rathjens, Jack Ruina, and Dean Wilkening all offered useful guidance. Herbert York granted me multiple interviews and shared his time generously. Lisbeth Gronlund and David Wright offered helpful comments and opened their home and files to me during the early stages of my research, as did John Kogut and Michael Weissman.

Members of the computing community have also offered extensive help. Herbert Lin offered invaluable assistance, including correspondence, interviews, and critical readership. Peter Neumann answered many questions and sharpened my arguments and prose. Barry Boehm, Danny Cohen, and Charles Seitz granted me interviews, and Jon Jacky, Greg Nelson, Brian Randell, David Redell, and Terry Winograd offered helpful correspondence and documents. In 2003 Dan McCracken shipped five large boxes of files to my office at MIT, in 2004 he allowed me to haul them out to Stanford, and during the final month of his life, he granted permission to donate them to the Charles Babbage Institute. I am especially indebted to Fred Brooks and David Parnas for correspondence that helped me more fully understand the context and content of their foundational work. Finally, archivists have offered me considerable guidance in my research, including Nora Murphy at MIT and Bill Burr at the National Security Archive.

I thank the many colleagues who helped shape my work through workshops, conferences, and other venues, including Janet Abbate, Nathan Ensmenger, Gary Downey, Thomas Haigh, Stephen Hilgartner, John Krige, Michael Lynch, Sean Malloy, Suzanne Moon, Gina Neff, Lisa Onaga, Megan Palmer, Ted Porter, Pablo Schyfter, and Matt Wisnioski. Many other scholars provided challenging questions and comments in response to presentations, at places such as Harvard's Belfer Center for Science and International Affairs, Cornell's Science and Technology Studies program, Virginia Tech's Science Technology and Society Program, the University of

British Columbia's Centre for International Relations, the history of science colloquium at UC-Berkeley and UC-San Francisco, and Stanford's Center for International Security and Cooperation.

I am thankful to my editors and reviewers for improving the concepts and prose of this book. Hugh Gusterson and Patrick McCray reviewed the book and offered many helpful suggestions and corrections. Katherine Goldgeier trimmed, lightened, and sharpened my prose. Marguerite Avery, Katie Persons, and Marcy Ross at the MIT Press also offered patient guidance in the finishing stages of this book. Finally, I am grateful for the friends and family who have cheered me on through the years. My mother, father, and brother have never wavered in their support. I am also thankful for the friendship of Pamela Hieronymi, Andy Kalt, and Salying Wong, who walked with me through the ups and downs of this book and much more. Finally, I am grateful for Doug MacMartin, who saw me through the final two years of writing this book. For his endless positivity, unwavering support, and generous love, I dedicate this book to him.

Introduction

Late on the night of March 22, 2003, a crew of U.S. Army soldiers pored over the green console of a Patriot missile defense battery in Iraq, watching for the symbols that warn of an enemy missile or aircraft. After hours of scrutiny, the sign for an anti-radiation missile appeared. The Patriot crew sprang into action, attempting to confirm the target before finally firing missile interceptors. A small pound sign appeared to confirm the kill. But soon, they learned they'd hit a British Tornado fighter plane. Its two crew members were killed instantly.

"It's obviously a software glitch," declared one defense official.[1] Within days, similar comments filled media commentary about the Tornado downing and similar accidents. But why was software such an "obvious" source of failure? Computers haven't always seemed so error-prone. Over 50 years ago, the first real-time computers were developed to help defend the United States against a nuclear attack, precisely because they were seen as the ultimate in speed and reliability. *Newsweek* described the system in 1958: "Lightening fast and unerringly, the electronic unit of the SAGE air-defense system supplemented the fallible, comparatively slow-reacting mind and hand of man."[2] Defense Department scientists soon spoke proudly of "a completely automated system to acquire, discriminate, track, fire, and command" missile defenses, "with no human entering into this chain."[3] Engineers, policymakers, and journalists easily imagined the earliest computers as fast, reliable aides or even replacements for slow, fallible human operators.

How did this change? Today the risks of complex software, as well as complex technology in general, come as no surprise to those familiar with nuclear reactor accidents and space shuttle crashes. We now have a rich literature showing how unexpected interactions in complex, fast-moving technological systems produce what sociologist Charles Perrow has called "normal accidents."[4] Complex weapons systems fail for even more fundamental reasons: an intelligent adversary is working to thwart the system; the threat that a system is designed to confront often changes by the time it is deployed.[5] At the same time, war planners often face pressure to quickly adapt a weapons system to a new operating environment.[6] Confronted by unpredictable and fast-changing environments, complex weapons systems are bound to fail at least some of the time.

But how did complex software come to exemplify these risks? Why have computers become such common suspects when complex technology fails? And if complex software poses such obvious risks, why does the United States continue to rely so heavily upon complex weapons systems, rushing untested technology into the field?

This book answers these questions by showing how some of the most complex weapons systems imaginable—missile defenses—came to be seen as prone to catastrophic computer failure. It tells the story of how an elite group of scientific advisors first grappled with the problem of defense and made their arguments politically persuasive, all the while neglecting the risk of catastrophic computer failure. It also shows how the rise of computing made these risks a focus of Defense Department advising and public debate.

The primary goal is to illuminate the messy process of constructing persuasive arguments about technology, ranging from informal memos to formal studies, from top-secret reviews to widely publicized articles. To do this, we must examine not only what was finally published or what became mainstream, but what was deleted or marginalized, and why. By looking inside the advisory process, this book also sheds new light on missile defense, a pivotal issue in nuclear policy. But it is not primarily about missile defense as a set of artifacts, institutions, or policies. Rather, this book uses missile defense to illuminate a broader set of processes: how scientific experts form judgments about complex technological proposals, how they make those judgments politically persuasive, and how a new field of knowledge becomes publicly authoritative. An understanding of these processes sheds light not only on missile defense, but on environmental climate models, genetically modified organisms, and similar complex systems that intertwine technological and political futures. How do scientists assess the future of such systems? How do they make their arguments politically persuasive? And how do these arguments change over time?

To clarify these questions, I introduce the notion of *disciplinary repertoires*—the quantitative rules, codified knowledge, and habits of problem solving that enable experts to structure, estimate, and quantify uncertain technological futures. Disciplinary repertoires allow experts to rhetorically distinguish subjective, politically controversial aspects of a problem from putatively objective, technical realities.[7]

But why disciplinary repertoires? Why not simply talk about disciplines? The answer is that disciplines are notoriously hard to define, shifting shape and form over time. Scholars in science and technology studies have thus focused not on the ideal type of disciplines, but instead on how knowledge is produced through diverse and changing practices.[8]

At the same time, if we only talk about disciplinary practices, then we miss an important aspect of knowledge-making: experts *perform*. As science and technology scholar Stephen Hilgartner has shown, expert advising can be seen as a dramaturgical display of scientific authority and objectivity.[9] Experts need to persuade not only their peers but also diverse public audiences that their claims are authoritative. The notion of a repertoire evokes a sense of practice, along with the dramatized, staged nature of so much expert advice today.[10]

Importantly, different disciplinary repertoires reveal different aspects of a problem. Just as a gestalt shift changes a rabbit into a duck, a new repertoire can radically alter scientists' perception of a complex technological problem.

For example, physics offers experts a means of analyzing how a complex system will work in the ideal world, without human actors. Though the cultures of physics have varied with time and place, in the twentieth century physics was often imagined as profound simplicity.[11] Albert Einstein, icon of 20th-century physics, reportedly stated that "everything should be made as simple as possible, but not simpler."[12] Physics draws attention to a world without social beings, a world constrained only by nature.

By contrast, computing offers experts a means of analyzing complex processes that are simultaneously social and technological. As it emerged in the Cold War, computer and software engineering offered humans ways to manage complexity. The best software, like any good engineering design, is elegant. Yet software inevitably involves what one computer expert, Fred Brooks, termed "arbitrary complexity": neither divine nor elegant, but "forced without rhyme or reason by the many human institutions and systems to which [software] interfaces must conform."[13] Computing draws attention to the messy, evolving relations between social organization and technological development and use.

Even today, the most prominent independent analyses of missile defense focus on the *physical* advantages of offense over defense. This critique has venerable roots. As the missile race heated up in the late 1950s, elite scientists and engineers were inventing myriad physical penetration aids— decoys, chaff, and other means of deceiving and overwhelming defenses. But in formal reports and public debate, scientific advisors rarely discussed an even more fundamental problem: even if all the defensive weapons and components work perfectly, the *informational* challenges of the complex, tightly coupled system rendered it prone to catastrophic failure. Without generally accepted, quantitative rules to clarify an uncertain future, they had neither the desire nor the ability to study the problem of software. In the late 1960s, discussion of complex software failure remained anecdotal rather than absolute, and was brushed aside as pessimism.

In the 1970s, that attitude began to change. Computer experts advocated "software engineering" as a means of managing the complex software, and sought to develop codified rules and practices. Some focused on management processes, developing quantitative rules to predict the time and costs of large software projects. Others focused on product reliability, documenting failures and developing means to prove that software was correct. Software engineering never became a well-defined scientific discipline or profession. It remains a motley mix, divided between managers and programmers, theorists and pragmatists, academics and industrialists. Nonetheless, a disciplinary repertoire enabled experts to analyze the problem of complex computer systems in the authoritative idiom of science.

The risk of catastrophic computer failure finally emerged as a central problem in the mid-1980s, after Ronald Reagan proposed that scientists and engineers build a missile "shield" that would render nuclear weapons "impotent and obsolete." Most scientists framed the Strategic Defense Initiative, known as "Star Wars," with basic physical principles. But a few began to draw from the new repertoire of software engineering. And when a prominent scientist resigned from a panel on Star Wars computing, citing "fundamental reasons" that the defensive software would never be trustworthy, the risk of catastrophic failure prompted a highly public debate. The U.S. Congress commissioned a study that concluded there was a "significant probability" that the Star Wars system would fail catastrophically if it were ever needed in a nuclear war.

By tracing the history of an often-marginalized argument, this book counters the common assumption that expert judgments are mirror reflections of technological realities. The goal here is not to argue that one form of argument—physical or informational—is superior to the other. Rather, it is to show how scientists construct judgments about complex, unpredictable weapons systems, and how those judgments gain traction in the political process—in other words, what makes those arguments count.

But how and why should we study often-neglected arguments? Isn't neglect a sign of irrelevance? No. In what follows, I briefly discuss the "how" of studying marginalized arguments, before delving into a fuller discussion of the "why." As we will see, this account carries significant implications, both for current debate about missile defense, and for our understanding of nuclear history. Additionally, this account illuminates several key processes of broader interest—how experts formulate distinctive conceptions of risk, how those conceptions change over time, how a new and contested field comes to be recognized as authoritative in the political arena, and how political culture shapes what counts as authoritative knowledge.

Studying a Marginalized Argument

This book began as a study of public controversy over President Ronald Reagan's Strategic Defense Initiative, or Star Wars. The nation's scientific elite engaged in often-heated discussion about the prospects for laser weapons and other exotic technology. Since I worked closely with lasers while earning a PhD in physical chemistry, these highly public debates fascinated me. As a postdoctoral scholar, freshly equipped with insights from the sociology of scientific knowledge, I set out to analyze the sources of disagreement and their political repercussions. But even as I scrutinized controversy over physical dimensions of the system—the number of lasers in space required to destroy the Soviet arsenal, the likely power of new laser weapons, the operational challenges of directed energy weapons, and the countermeasures that might render exotic weapons ineffective—I concluded that the most compelling technical arguments against Star Wars were not buried in these arcane calculations.[14]

In fact, some of the most persuasive technical arguments against the Strategic Defense Initiative only emerged publicly after two years of highly public debate, when a prominent software engineer resigned from a government panel on computing. Why, I wondered, did it take so long for computing to become a subject of public debate?

The answer might seem obvious: the most prominent critics of SDI were physicists with long experience advising the U.S. government. But a closer look complicates matters. The first quantitative critique of SDI's computer system came from Herbert Lin, a young physicist who used a textbook on software engineering. Richard Garwin, one of the most elite physicists to critique SDI, was a senior scientist at IBM, but he hardly mentioned computing challenges. Neither disciplinary training nor institutional affiliations determined the way that experts analyzed problems. History mattered, but in a more complicated way than one might first think.

Digging into the history of scientific advising on missile defense and nuclear policy, I searched for clues about how experts viewed the risk of failure in complex, computerized systems, and how those views changed over time. This meant digging into the archives of well-known physicist-advisors, as well as computer-related archives. Useful sources include papers by computer experts who, while well-known in their own field, would never appear on a "who's who" list of prominent public intellectuals. Congressional records, the National Archives, and the National Security Archives at George Washington University have also provided rich detail and a wealth of now-declassified reports. Interviews with noted physicists and computer experts have also been invaluable. While I have hunted for what experts did say about missile defense, I have equally sought to understand what they did *not* say, and what was *not* taken up in the political process.

It may seem odd to go looking for advice not given or heard. When I asked Garwin why he never discussed software, he replied: "All my life, I have been extremely busy, and I don't do any more in experiments or in analysis than is required to arrive at a definitive answer. So when I find that a system has a fatal flaw, I am not particularly interested in extolling other virtues of the system or in finding other flaws."[15] Garwin's reasoning was simple: "the systems would fail for physical reasons, so why should I waste my time on software?"[16]

Why, indeed? This answer, from such an eminent scientist, might seem to undermine the very purpose of this book. But in fact it suggests an important part of the argument. Scientific advisors working with limited resources, usually on a part-time basis, use the intellectual tools closest at hand. They aim at the minimum analysis necessary to persuade decision makers to take some course of action. Without software engineering—a disciplinary repertoire that only began to emerge in the 1970s—experts were not inclined to "waste time" on software. Nonetheless, the marginality of arguments about computing and software carry consequences, both for our understanding of nuclear history, and for contemporary policy.

The Consequences of Forgotten Arguments

A close look at the historical record shows that concerns about computing for missile defense were surprisingly persistent, yet remained marginal for decades. As early as 1963, the well-known physicist Herbert York argued that offense had a great advantage over defense, "the advantage of people working many years to try to develop penetration aids over a computer which must solve the problem in a matter of a few minutes."[17] Opposing missile defense deployments publicly in 1969 and 1970, York argued that missile defenses were likely to fail catastrophically the first time they were really needed. Computer experts made similar claims at the time.

These arguments are typically overlooked by scholars who focus on the *outcome* rather than the *process* of expert advising. Although scholars and journalists have examined the role of experts in missile defense controversy, none has looked closely at *how* scientists and engineers formed judgments about missile defense, or how those arguments became influential in the political process.[18] Some studies neglect divisions among scientists, treating technological capabilities as a "pure" social construction, or as self-evident facts.[19] When they do examine controversy, most studies tend to take sides, presenting the views of some scientists as authoritative, while casting others as biased.[20] Either way, most accounts of missile defense tend to treat scientific advising as something that is ideally insulated from politics—a rational input to be selected, rejected, or simply ignored.[21]

By contrast, this book joins works emphasizing that it is impossible to find a principled boundary between scientific advising and political decision making.[22] This is not to say that there are not more or less objective positions in any analysis, but rather that there can be no "purely" scientific advice on missile defense. As we will see, missile defenses exemplify what Gabrielle Hecht has termed "technopolitics," meaning their technical design reflects and enforces specific political goals.[23] The components of defensive systems must be geographically distributed in ways that require cooperation from diverse populations. Analysts cannot calculate the technical effectiveness of defense without making assumptions about the future offenses of political and strategic adversaries. Furthermore, experts endeavor to make their arguments persuasive in the bureaucratic, presidential, and legislative processes associated with defense policymaking.

The fact that some arguments "stick" in these processes, while others do not, carries consequences. In the early 1980s, scholars and defense analysts suddenly began producing a deluge of articles, books, and reports analyzing the limits and risks of complex, computerized nuclear operations.[24] These risks might have been minimized if they had become a focus of analysis two decades earlier—for example, if the 1960s Defense Department had been dissuaded from developing technologies such as Multiple Independent Reentry Vehicles, which dramatically shortened times for decision making, and increased reliance on computers. The conventional explanation for such developments emphasizes bureaucratic politics, and the tendency for services to compete with one another for the biggest, most glamorous weapons.[25] But if we stop here, we fail to appreciate why defense department organizations privileged some solutions to the challenges of nuclear weapons development, while neglecting others.[26] As we will see, Defense

Department analysts were focused on quantitative physical and economic measurements of weapons effectiveness, and had no way of quantitatively or authoritatively analyzing the risks of failure in complex computer systems in the 1960s. As a result these risks were too easily pushed aside in the defense procurement process.

Arguments about the risks of complex software remain salient to today's efforts to deploy defenses that will be both "proven" and "adaptive," giving the United States the "flexibility" to rapidly counter emerging threats.[27] Many physicists have argued that defenses are not proven, pointing to failed tests and the fact that not even "successful" tests have realistically represented the likely battle conditions. The defense has known precisely when the test target was to be fired, and the environmental conditions and target were deliberately chosen for easy interception.[28]

While scrutiny of the physical testing program is valuable, an analysis of the challenges of complex software gives us reason to question whether any amount of field testing can yield defenses that are *both* proven and adaptive. Software is limited by its arbitrary complexity—the requirement that it conform to many human institutions that change frequently and unpredictably. For example, the computerized command and control systems of U.S. missile defenses must interact not only with U.S. defensive radars and weapons, but also with the command systems and technologies of allies, all of which change over time. As a result, software that is proven in one context is not trustworthy when it is first adapted to new conditions—it only becomes reliable to the extent that it sees operational use and debugging in a highly specific, arbitrarily complex set of conditions. No matter how physically realistic tests of experimental defenses become, or how well they perform in a "real-world" conditions, defensive software cannot be *both* proven *and* adaptive to new conditions.[29] Indeed, the challenge of integrating many different command and control systems has recently drawn attention from government investigators concerned about the risks of U.S. efforts to deploy defenses in Europe.[30]

A Political History of Computing

By following the changing place of software in a high-stakes debate, we also gain a different kind of computing history. Historians of computing have been increasingly attuned to politics in recent years. This study joins several key works in emphasizing the entanglement of technology, its representations, and political power. Computer historian Paul Edwards has argued that real-time digital computers embodied the military's interest in controlling nuclear conflict.[31] Yet this meaning was not fixed. As communication scholar Fred Turner has shown, a few key entrepreneurs and journalists sought to transform computing in the popular imagination, from representing a "closed world" to symbolizing countercultural and economic freedom.[32] While these accounts centralize the persuasive skills of individual entrepreneurs—engineers, journalists, and businessmen—we know relatively little about how computer experts *qua experts*, as a social group, have shaped the popular meaning of computers.

There are some good reasons for this neglect, for historians have found the field of computer expertise notoriously difficult to define. Does computing qualify as a "profession" or "science"? Classical sociological measures offer little guidance. Does professionalization mean the right to control who can do certain kinds of work, as suggested by Andrew Abbot?[33] Computing fails this test; many programmers possess no special credentials. Does professionalism follow from an applied "science"?[34] University computer science programs have burgeoned in the past fifty years, but academic "computer science" never became a criterion for hiring in industry.[35] In fact, the most innovative work in computing emerged not from deliberate efforts at discipline-building, but through the innovative work that took place at the intersection of multiple and sometimes conflicting institutional interests.[36] Indeed, computing involves so many different kinds of work that it makes little or no sense to speak of a computing profession.[37]

While the heterogeneous field of computing challenges traditional notions of the sciences or professions, it is also paradigmatic of many fields—such as synthetic biology, nanotechnology, or materials science. Indeed, as sociologists since Thomas Gieryn have emphasized, most "disciplines" are bounded more by rhetoric than by practice.[38] Thus, rather than addressing questions of how or when computing became a proper "profession" or "science," this account focuses on how computer experts have negotiated these charged categories over time. In other words, I focus on a practical problem: how do computer experts *perform* authoritatively as professionals or scientists?

I argue that "software engineering" emerged as a solution to a problem that was simultaneously technological and performance-based. From the very earliest computer projects, engineers struggled to understand and represent "software." Was it research, development, or production? In practice, software tended to conflate all three, leading to conflicting expectations and repeated fiascoes. When software experts acknowledged this conflation in the late 1960s, they blamed it on social pressures, and sought to establish the new field of "software engineering" as a means of gaining mastery.

But by the 1980s, many had come to view the entanglement of research, development, and production as a fundamental aspect of software technology. For these experts, the disciplinary repertoire of "software engineering" became a tool for speaking authoritatively about the challenges of complex software. It enabled them to mark claims about risky systems as "technical," rather than "political."

By showing how a disciplinary repertoire enabled computer experts to resolve the practical problem of performing authoritatively, this book extends scholarship on the *practices* of making new knowledge. While much work on the practices of science has focused on the contexts of laboratories and classrooms, we will see that disciplinary repertoires also shape knowledge-making in the politically-charged arenas of government advising, deliberation, and controversy.[39]

When and Why Do Arguments Count?

While disciplinary repertoires may enable experts to project an air of scientific authority, they do not necessarily make an expert persuasive. Ultimately, persuasion, credibility, and authority are *relational*—that is, they are the product of a relationship between a performer and his or her audience. What pleases one crowd may leave another cold. Experts that impress one group may evoke skepticism from another. One goal of this book is to show how *different kinds of arguments* become persuasive in relation to particular contexts of policymaking, and how this in turn shapes what counts as authoritative public knowledge.

Scholars in science and technology studies have demonstrated that the authority of experts is shaped by political culture—the symbolic practices by which public decisions are made and remade.[40] For example, in the United States, fears of technocracy prompt the maxim that science and politics should not mix. We do not want scientists to be making decisions for a nation whose collective will they may not represent. Neither do we want scientific truth to be skewed by politicians vying for power. Yet in practice, established scientific knowledge leaves considerable uncertainty about appropriate policies. In fact, a few scholars have gone so far as to claim that traditional disciplines are too specialized to address real-world problems that entangle technology and politics, and hence decisions about such problems reduce to "pure" politics.[41]

More commonly, scholars in science and technology studies acknowledge the importance of specialized knowledge to sound policymaking, but emphasize the need for negotiations between experts and elected officials.[42] For example, institutions such as the National Academies of Science, the Food and Drug Administration, and the Environmental Protection Agency produce authoritative claims by carefully negotiating the relationship between scientists and policymakers, while publicly emphasizing the independence of these groups.[43] Strong fears of centralized power have created an elaborate system of checks and balances, encouraging adversarial proceedings. As science studies scholars such as Sheila Jasanoff and Steven Epstein have shown, public controversies themselves contribute to knowledge-making.[44]

Yet, none of these studies has closely examined how specific *kinds* of knowledge become persuasive in the diverse contexts of policymaking. When and why have the arcane calculations of theoretical physicists proven more persuasive than the "common-sense" arguments of software experts? When has the opposite been true? By comparing how different kinds of knowledge have proven persuasive, in multiple decision-making contexts, this book highlights the relationship between disciplinary and political cultures.

Arcane, quantitative calculations are the stuff of modern bureaucracies. As we will see, idealized calculations—especially the projected costs and physical effectiveness—quickly became a dominant means of analyzing missile defense in the highly secretive deliberations of the Defense Department. However, the rules of the game change when an internal bureaucratic debate becomes a matter of highly visible, public controversy. Calculations do not make good talking points for a general audience; instead, scientists gained public salience by developing qualitative arguments that aligned with dominant political agendas, and by using windows of political opportunity.[45] Concerns about relatively unpredictable and nonquantifiable computer failures have the "common-sense" appeal of Murphy's Law—the adage that if something can go wrong, it will—and thus have historically emerged during periods of heated public debate about missile defense.

Nonetheless, arguments about the intrinsic fallibility of technology often face an uphill battle in the United States. As historian Thomas Hughes has argued, high technology is a symbol of national identity in the United States.[46] As a result, technology policy is often determined not by careful government regulation, but by de facto commitments to free markets.[47] Faith in technological progress has been institutionalized in modern corporations, their publicity, their workforces, and their markets. Engineers who acknowledge limits to technology are all too easily cast as pessimists or failures. Policymakers who express caution about prospects for technology put themselves in a vulnerable position. As a result, critics of new technology face pressure not only to show that a proposal is *probably* flawed, but that it is *provably* flawed.

This book shows how scientists have attempted to prove such limits to missile defenses, using distinctive disciplinary repertoires, and how these arguments have fared in political debate. Importantly, physics suggests attention to the limits of "nature" rather than "technology." Such arguments seem to enable proofs. By contrast, the limits analyzed by computer experts reside in complex, human-built technologies. Arguments about such limits tend to take more probabilistic form, and do not fare well in a culture of technological enthusiasm.

The Road Ahead

The organization of this book underscores a key disjuncture between two communities: a highly visible circle of scientists and engineers (mostly physicists) who served as elite advisors to the president, and an eclectic group of scientists and engineers who converged around computing technology. The chapters progress roughly chronologically and are punctuated by significant political developments, but the focus shifts between these two communities, with odd chapters focusing on technologists, and even chapters highlighting elite deliberations, for most of the book.

Chapter 1 traces the disjuncture between elites and technologists to wartime research and development, and shows how the different ways in which these groups engaged with air defenses led to radically different understandings of computer programming in the 1950s. Drawing on newly available archival materials, it explains why programming—recognized by historians and participants as the "most underestimated task" in air defense—was initially imagined as a trivial task.[48] Because scientists framed computerized air defense with the disciplinary tools of physics and electrical engineering, they viewed programmability as a special advantage of digital computers, something that would make computing machinery "flexible" and adaptable to changing threats. Elite scientists endorsed computerized air defenses with this thin understanding of computers, and remained focused on their primary goal: shaping military policy. But the technologists who were tasked with developing computers for air defense were soon grappling with managing the most complex computer programs ever developed. Technologists discovered that programming challenged traditional boundaries between engineering "research" and "production." Many came to find research challenges in the "black art" or nascent science surrounding systems programming. But the advisory elite viewed programming as a "production-oriented" activity, not as a field requiring research.

Chapter 2 shows how the elite's tendency to black-box computer programming influenced their framing of an "appallingly complex" problem—defense against intercontinental ballistic missiles—in the late 1950s. While past accounts have highlighted scientific opposition to deployment, in fact scientists did not so much oppose defensive deployments as reorient them toward more limited goals that might contribute to arms control.[49] In particular, they hoped that by protecting against an unexpected or accidental launch or small attack, defenses might facilitate an agreement to limit offensive weapons. Significantly, such defenses would depend critically upon complex computer programs, yet existing evidence suggests that scientific advisors never expressed concern about the reliability of software. On the contrary, because they recognized the need to respond rapidly to the threat of ICBMs, they endorsed the computerization of early warning, intelligence, and other military command and control systems. They treated computers as devices that could be readily produced by industry, and did not seem to anticipate any challenges associated with programming such systems.

So how did elite advisors come to understand software as a field of scientific research, and how did this shape their analyses of missile defense? Chapter 3 tackles these questions, showing how the military's pursuit of a plethora of incompatible and costly computerization projects posed a management dilemma for a cost-minded Kennedy administration in the early 1960s. The Defense Department partly addressed the problem through a new Information Processing Technologies Office. While this office and its charismatic first director, J. C. R. Licklider, have been well-studied for their influence on the field of computing, we have yet to understand the relation between Defense Department sponsorship of computing and its parallel programs in missile defense.[50] As we will see, Licklider framed "the software problem" in terms of human-computer interactions and thus gained support from military officers concerned about command and control, as well as from scientists interested in using computers more effectively. But since elite physicist-advisors expected that missile defense would be completely automated—there would be no time for humans to get into the decision-making loop—they saw no "software problem" for missile defense. Instead, they continued to draw upon their primary disciplinary repertoire—physics—to advise the government about defense.

Chapter 4 shows how the same scientists who sought to improve missile defenses came to argue that the United States and the Soviet Union should agree to *never* deploy defenses, and how they drew upon their disciplinary repertoire to make persuasive public arguments. When the Johnson administration decided to deploy the so-called Sentinel missile defense in 1967, physics became a means of turning inside knowledge into persuasive public claims. Physicists' widely cited critiques helped provide the U.S. Congress with a way to challenge the president's military policies in the late 1960s, amid growing protests against the Vietnam War. The new publicity also opened the door to new kinds of arguments about missile defense. Physicist Herbert York spoke not on the basis of formal analysis, but on the basis of common sense, when he argued that the tremendous complexity of missile defenses would make them prone to "catastrophic failure."

York's remarks opened a public debate about computer fallibility, but they offered little in the way of systematic analysis. Chapter 5 shows how computer experts attempted to speak authoritatively about the risks of software for the "Safeguard" missile defense system (the successor to Sentinel), just months after the birth of "software engineering." It examines the origins and influence of a little-studied movement—Computer Professionals against the ABM (antiballistic missile)—showing how its professional anxieties and lack of formalized knowledge limited its intervention in the public debate. These limitations were evident in the efforts of Licklider, one of the best-connected members of the computer professionals' movement. Expressing anxiety about "anecdotal evidence," Licklider felt that he could only urge "a subjective appreciation" of the challenges of developing complex software. And without consensus on these challenges, many senators insisted that engineers would overcome the difficulties with sufficient effort. The U.S. Senate approved deployment of defenses around Minuteman silos by just one vote in August of 1969, and deployment began a few months later.

So did arguments about software actually shape decision making about missile defense? Yes—but in a subtle way. Declassified documents show how concerns about software began to shape debate about missile defense—and with it the secret, duplicitous negotiation of the Anti-Ballistic Missile Treaty in 1972. Chapter 6 uses these documents to reveal the mutual shaping of technological designs and political controversy.[51] The complex, geographically distributed nature of missile defense rendered it vulnerable to political controversy, even as political dispute exacerbated the challenges of developing complex software. As scientists warned policymakers about the challenges of complex software, they not only encouraged the negotiation of the ABM treaty, but also influenced the long-term direction of the U.S. missile defense program.

The public missile defense debate was just one aspect of a broader transformation in the political environment surrounding defense research and development, one that nurtured software engineering research. Several historians have analyzed the world's first "software engineering" conference in 1968 and its ill-tempered successor in 1969, but none has yet considered what came next: in the late 1970s, after a six-year hiatus, software experts began to establish a disciplinary repertoire around software engineering.[52] Chapter 7 shows how a few influential software experts argued that software occupied a growing fraction of computing budgets, and advocated a solution: investment in software engineering research. In a process similar to what sociologist Geoffrey Bowker has termed "legitimacy exchange," software engineering gained institutional status, as experts projected a sense of control over software development.[53] Practitioners never agreed upon the boundaries of the field, and clashes were frequent, between managers and technicians, academics and industrialists, pragmatists and purists. Nonetheless, as experts conversed and networked at conferences, in research journals, and in special-interest groups, they established a disciplinary repertoire—a heterogeneous set of principles, quantitative rules, and documented experience.

How did this new disciplinary repertoire shape debate about President Ronald Reagan's 1983 proposal to build a defense that would "render nuclear weapons impotent and obsolete"? As chapter 8 shows, the new repertoire influenced the ways that well-established scientific advisors analyzed the Strategic Defense Initiative (SDI), or "Star Wars" missile defense program. While most accounts of SDI have focused on the role of physicists in the political debate, this account reveals the growing influence of computing.[54] The first study of SDI computing placed highest priority on the challenge of designing battle-management software, calling it "a task that far exceeds the complexity and difficulty that has yet been accomplished in the production of civil or military software systems." Although elite physicists had relatively little to say about computing, a few younger physicists began to analyze the challenges of software, drawing upon the disciplinary repertoire of software engineering. And when a prominent software engineer, David Parnas, resigned from a Defense Department panel on SDI computing, the feasibility of software became a subject of nationwide debate. Ultimately, the U.S. Congress commissioned a study by the Office of Technology Assessment, which concluded that there was a "significant probability" of catastrophic failure in a missile defense system.

The pursuit of missile defenses continues today. A concluding chapter shows how the risks of complex software systems have continued to surface in the missile defense debate since the end of the Cold War. As we will see, software engineering experience carries significant implications for contemporary missile defense policy. This experience dates back to the beginning of the nuclear age, when the threat of attack from Soviet bomber planes prompted scientists and engineers to build the first digital, electronic, real-time computer systems.

1 Software and the Race against Surprise Attack

The computer programming was probably the most underestimated task in the entire SAGE project.[1]

—John Jacobs, 1983

It seemed to come together by a chance conversation. One chilly January day in 1950, Jerome Wiesner happened to see George Valley in the hall at MIT, asked what was new, and got an earful. Just a few months earlier, the Soviet Union had detonated its first nuclear weapon, catching many scientists by surprise. Valley, a physicist, recognized that a Soviet attack was all too easy. By flying under long-range radars, a single bomber could destroy an American city before the military could begin to mobilize. Hundreds of shorter-range radars might detect low-flying bombers, but this wouldn't leave enough time to respond. By the time military officers started the complex, laborious work of gathering data and calculating the trajectories of hundreds of aircraft, bombs would be flattening cities.[2]

Valley knew that in principle, a digital electronic computer could track the aircraft in real time. But he could not find what he needed. Most computers were essentially giant calculators, not machines that could keep up with fast-moving events in the real world. Wiesner told Valley that the computer he was envisioning was not only under development, it was "up for grabs" at MIT.[3]

One week later, Wiesner and Valley lunched with Jay Forrester, an electrical engineer struggling to build the world's first real-time digital computer. Soon Valley and dozens of engineers were designing a computer to integrate air defenses. A complicated array of radars, telephone lines, radio stations, fighter planes, antiaircraft guns, and missiles would all be coordinated by computers in command centers, the Semi-Automatic Ground Environment (SAGE). Humans could interact in real time with the machines, but computers would be the brains of SAGE, and the lynchpins in a nationwide defense against nuclear attack.[4]

But when it came to actually building the system, the SAGE engineers fell into an unexpected quagmire: specifying, writing, testing, and taming complex software. Years later they recalled the early "myths": that they could "do anything with software on a general-purpose computer; that software is easy to write, test, and maintain; that it is easily replicated, doesn't wear out, and is not subject to transient errors. We had a lot to learn."[5]

Virtually every history of the SAGE air defense effort has noted that the difficulties of programming caught engineers off guard.[6] But why were the SAGE engineers so surprised? Why did they ever imagine that it would be *easy* to write computer programs that could integrate myriads of disparate electronics into a single, coordinated system? And how did their hard-earned lessons take root among the scientists and engineers who became the nation's elite advisors on national security?

The answer lies in the intellectual and social roots of the air defense effort. This chapter describes an early rift between physicists such as Wiesner, who became part of an advisory elite after World War II, and the scientists and engineers who developed the new fields of computing and systems programming. Initially, both groups imagined that programming the air defense computers would be easy, in large part because they designed the computerized air defense system with the disciplinary repertoires of electrical engineering, physics, and mathematics. They imagined computer programming as something that would be far more flexible and versatile than physical electronics. Furthermore, the technologists developing SAGE had every incentive to portray programmability as a special advantage of digital computers; they needed to persuade more elite scientist-administrators and policymakers of the value of their research.

But as they dug into the programming job, SAGE engineers found themselves scrambling to hire and train hundreds of programmers, managing an unprecedentedly complex programming project, and struggling with their professional identity. Programming challenged traditional distinctions between "research" and "production." The engineering research used to develop program specifications could not be separated from the organizational challenge of managing program development and debugging. Some technologists came to find interesting challenges in the problems of developing and managing complex computer programs, they began to forge a new field of systems programming.

However, systems programming was not recognized as a science by the nation's scientific elite. The engineers who learned the hard lessons of SAGE programming never belonged in the most elite ranks of physicists advising the government. By the time the technologists had learned the hard lessons about programming, the more elite physicists had moved on to a new problem, at least on paper: missile defense.

1   The Social Roots of SAGE

1.1   Inventing Programming

During World War II, most scientists and engineers used the term "computers" to describe relatively low-paid workers, typically women, who used calculators to compute anything from the implosion mechanism for the atomic bomb, to ballistics tables that gunners could use for fire control. The architect of wartime research policy, Vannevar Bush, had pioneered one of the most advanced calculating machines, the differential analyzer, which calculated by forming a model, or analog, of the physical world. During World War II, Bush's Office of Scientific Research and Development funded a variety of efforts to improve machines for computation, recognizing that scientists and engineers needed to "get the numbers out."[7]

At the University of Pennsylvania, the physicist John Mauchly and electrical engineer J. Presper Eckert proposed to develop the Electronic Numerical Integrator and Calculator (ENIAC), a machine that would work faster than other devices because it would be completely electronic. However, their initial proposal seemed too far-fetched to attract support from Bush's wartime research establishment. Instead, with the help of Army lieutenant and mathematics professor Herman Goldstine, they obtained Army funding. And with the help of the renowned mathematician, John von Neumann, the ENIAC spurred the development of the first stored-program computer.[8]

Originally a theoretical mathematician, the Hungarian-born von Neumann lost his "purity" when he began helping the military with computational problems during World War II, developing what he described as "an obscene interest in computational techniques."[9] He became one of the most respected and influential scientists in the defense advisory community during and after the war. In 1944, von Neumann discovered the ENIAC project through a chance encounter with Goldstine, and was immediately interested.

As Jennifer Light has discussed, the programming of the ENIAC was a conceptually and physically laborious job. To program the machine for a computation, a group of women (known as "the ENIAC girls") connected cables between different circuits in the machine.[10] Goldstine and von Neumann soon proposed an improved machine that would store the program routine in coded form. As described in a widely circulated report by von

Neumann and Goldstine, this new computer architecture inspired computer developers around the world who began to move toward von Neumann's ideal.[11]

The invention of electronically stored computer programs changed the work associated with calculation. The term *computer* came to apply to stored-program machines rather than their human operators, while a new field of work began to emerge around developing and coding such programs. With the help of mathematician Adele Goldstine, Herman's wife and the leader of the women who programmed the ENIAC, von Neumann and Goldstine wrote a volume that envisioned the emergent field as a subfield of mathematics. Von Neumann and Goldstine emphasized that "coding is not a static process of translation, but rather the technique of providing a dynamic background to control the automatic evolution of a meaning"; it was a "logical problem." They aimed to show how it could be "mastered" through "a new branch of formal logics."[12]

In practice, coding was a tedious, error-prone process that involved far more than logic—it required a complicated assemblage of machinery. Programmers might start with a flowchart on a piece of paper, a set of easy-to-read instructions showing what the computing machinery should do, and when—adding, subtracting, transferring numbers into memory, and so on. But translating these high-level instructions into machine code—binary strings of 0's and 1's represented on a piece of tape—could be quite tedious. After coding, the tape would then be fed into a photoelectric reader, which would generate binary electronic signals for the computer's use. After the computer finished its calculation, a teletype machine would transfer the electronic signals back to paper. After perhaps a few hours of waiting, the programmer could review his or her results to determine whether the calculation performed correctly—and if there was an error, he or she would repeat the process.[13]

In 1949, when researchers at the University of Cambridge achieved the first stored-program electronic computer—the Electronic Digital Scientific Analyzer and Calculator (EDSAC)—they were surprised by the sheer work of programming. Years later, EDSAC designer Maurice Wilkes remembered when the programming challenge came on him with "full force." The EDSAC was one floor up from the equipment necessary to write or change a program. Wilkes repeatedly amended code on the first floor and tested code on the second: "On one of my journeys . . . 'hesitating at the angles of stairs' the realization came over me with full force that a good part of the remainder of my life was going to be spent in finding errors in my own programs."[14]

Wilkes and others took steps to make programming more intuitive. Rather than writing all their programs in binary machine language, Wilkes introduced equipment that translated letter instructions into binary code. And his team designed "subroutine libraries," collections of short, frequently used programs stored on tape. These could be duplicated using special machines, allowing them to be incorporated into new programs without mistakes. Wilkes published his programming techniques in 1951, and they quickly became the standard in the field.[15]

Nonetheless, programming remained tedious. In most computer groups, programmers were not allowed to operate the computers, and hence could not quickly search for mistakes. Instead, computer users filed to a digital computer, where they handed their program tapes to a machine operator, and waited for the results to return. By December 1951, the first Joint Computing Conference in Philadelphia devoted special attention to the tedious and slow nature of programming.[16]

1.2   From the Rad Lab to Air Defense

Why then, were the scientists and engineers who developed the SAGE system surprised by the difficulty of programming? To answer this question, we must step back briefly to consider the social and political roots of the air defense project. The scientists and engineers who became most influential in national policy circles after World War II were not computer technologists. Instead, they gained status through wartime work on prestigious projects such as the atomic bomb and microwave radar.

At MIT, microwave radar developments spurred the establishment of the Radiation Laboratory in 1940, which drew top physicists from around the nation, and served as a training ground for a postwar scientific elite. Lee DuBridge, a physics professor and dean at the University of Rochester, became the lab's first director, and recruited his associate director, the eminent physicist F. Wheeler Loomis, from the University of Illinois. The prominent physicist Isidor Rabi took a leave from his professorship at Columbia, and eventually became DuBridge's associate director. Jerrold Zacharias followed Rabi from Columbia, taking charge of the Rad Lab's division on transmitter components, and was succeeded by one of DuBridge's former students, Albert Hill. Julius Stratton, a physics professor at MIT, joined the theory group.[17]

Many of the scientists who worked on microwave radar rose rapidly through MIT's administration after the war, and gained influence in Washington. Stratton became the director of a new Research Laboratory of Electronics (RLE) after the war, which inherited much of the equipment of the wartime Rad Lab. When James Killian became MIT's president in 1948, Stratton became MIT's first provost, and Killian's right-hand man.[18] Hill moved up from his position as the associate director to become its full director of the RLE. Jerome Wiesner, who worked under Zacharias and Hill at the Rad Lab, became the RLE's associate director before even earning his PhD in 1950. In the fall of 1945, the MIT administration charged Jerrold Zacharias with a new Laboratory of Nuclear Science and Engineering, which was modeled after the RLE, and recruited top physicists from the Manhattan Project.[19]

Although microwave radar was MIT's most prestigious project, it only became useful in the field through systems engineering. For example, the physicist Louis Ridenour who earned a PhD at Cal Tech in 1938, left a professorship at Princeton to became the director of the Rad Lab's airborne "fire-control" division—an effort to use electromechanical feedback devices (servomechanisms) to steer radar and guns, tracking objects automatically. Ivan Getting, who earned a PhD in astrophysics at Oxford in 1935, left a position at Harvard to direct the Rad Lab's division on ground-based fire control. George Valley was only two years out of doctoral work under DuBridge when he joined Getting's group in 1941.[20]

Like the physicists who worked on microwave radar, those who worked on systems engineering emerged from World War II with expanded levels of influence. Valley and Getting both became members of MIT's faculty, while Loomis recruited Ridenour to the University of Illinois. But while the physicists who worked in microwave radar were rising rapidly through academic administration, the systems engineers were finding influence in the Air Force, a new and growing service. Getting, Ridenour, and Valley all served on the Air Force's Scientific Advisory Board (SAB), founded in 1946.[21]

As a member of the Air Force's SAB when the Soviets detonated their first atomic bomb in 1949, Valley grew concerned about U.S. vulnerabilities to surprise attack. The air defense system was a patchwork of outdated World War II radars manned by Army officers in Quonset huts, communicating with a weak squad of Air Force fighters, using low-frequency radio sets that were as reliable as the weather. Furthermore, the radars could not see over the earth's horizon. This meant that low-flying bombers could evade radar until it was too late to respond. In November 1949, Valley proposed that the Scientific Advisory Board convene a panel to study the problems of air defense.[22]

The missive hit its mark. The Air Force had been focused on upgrading its bomber force, the Strategic Air Command. But in the wake of the first Soviet atomic bomb, Joe I, leaders felt increased public pressures to focus on defenses. And though the Air Force would continue to emphasize deterrence, even this required defenses. Operating doctrines required the Strategic Air Command to be launched only upon warning from defensive radars.[23] By December, General Muir S. Fairchild, the Air Force's vice chief of staff, asked Valley to accept chairmanship of the Air Defense System Engineering Committee (ADSEC), a group that included several members of the Scientific Advisory Board and MIT faculty.[24]

As Valley's group began to study the problem, they quickly concluded that an electronic digital computer was the only machine that could track the aircraft in real time. At first they could not find the kind of machine they needed. The first digital electronic computers were like giant calculators; they could not interact with fast-moving events in the real world. Valley needed a machine more like the real-time fire control systems of World War II.[25]

Fortunately for Valley's group, a real-time digital computer, the Whirlwind, was being developed at MIT. Jay Forrester, an ambitious graduate student in MIT's servo lab, had begun work on a general-purpose flight simulator for the Navy in 1944. After the war, Forrester and his fellow grad student, Robert Everett, slowly shifted plans toward a real-time, general-purpose, digital computer. Forrester and Everett were drawn to digital machines partly because they would reduce the accumulation of electronic noise in analog hardware. Additionally, digital machines were viewed as "much more versatile" than analog machines, because they could be reprogrammed for various tasks.[26] In 1948, Forrester's group outlined an expansive vision of what real-time digital computers could do for nearly every aspect of military operations: air traffic control, fire control, combat information centers on ships, antiballistic missile defense, and general command systems.[27]

1.3   Controlling the Whirlwind

When Valley discovered the existence of Whirlwind, he proposed a more focused, near-term goal for Forrester: a demonstration of how computers might improve air defense. However, in so doing he entered a dispute over the appropriate place of academic scientists and engineers in the postwar research establishment.

In the late 1940s, the elite administrator scientists of MIT—Stratton and Wiesner—embraced military funding, but sought to turn it toward basic science, rather than applied engineering projects like Whirlwind. They noted that military support for applied engineering could shift, while industry might develop similar projects, leaving MIT in a losing competition.[28] Indeed, by 1950, the Navy was growing weary of Forrester's plans, which seemed to continue expanding into the indefinite future, without ever producing a working computer. Most of the elites viewed computers not as a new field of research, but as tools for more established science and engineering research. Accordingly, in February 1950, Stratton began planning to make the Whirlwind part of a computation center that would serve scientists and engineers throughout the university.[29]

These plans were complicated by Valley's discovery of Whirlwind. Valley arranged to attend a meeting between Stratton and other MIT administrators and representatives from the Office of Naval Research in March 1950. There he proposed that the Air Force be allowed to partly fund Whirlwind for the study of air defense. Both MIT and the Navy agreed, and over the next several months, Valley's committee began to use Whirlwind to experiment with air defense concepts. By September 1950, they successfully demonstrated that the Whirlwind could receive radar data from a phone line and display it on a cathode-ray tube. Valley felt this was "proof of principle," just "sufficient for ADSEC's purpose, which was to tell the Air Force what to do, not to actually do it for them."[30] He envisioned writing a final report and getting back to life as usual.

But by that time the Korean War had started, and physicists were being asked to do much more than paper studies. Three former division heads in the Rad Lab, including Ridenour and Getting, had gone to work for the Air Force. Valley recalled that in October 1950, these men "demanded, and received a demonstration of ADSEC's 'achievement' . . ."[31]Whirlwind's very crude display of radar data was persuasive. They wanted to establish an air defense laboratory at MIT, akin to the "Rad Lab," but under Air Force sponsorship. Ridenour soon approached MIT's provost, electrical engineering professor Julius ("Jay") Stratton, describing Valley's work as "our brightest hope" in air defense, and suggesting that MIT establish a laboratory.[32] Valley recalled that in December 1950, Ridenour "coaxed" him to draft a letter proposing that MIT establish an air defense laboratory.[33] By 4 p.m. that day, it had been signed by Vandenberg and sent to MIT president James R. Killian.

Killian and Stratton were wary of the proposal. They sought to avoid commitments to such applied engineering projects, instead supporting national security through advising. Indeed, Stratton encouraged Zacharias, Hill, Wiesner, and others whom he viewed as MIT's "extraordinarily gifted group of leaders" to take on "ad hoc" studies for the military.[34] For example, in early 1950, Zacharias led a Navy-sponsored study of submarine warfare, Project Hartwell. In studies like Hartwell, civilian scientists enjoyed broad discretion in organizing studies and making recommendations. As we shall see, these broad ranging studies paved the way for physicists like Zacharias and Wiesner to become presidential science advisors.[35]

However, neither Valley nor Forrester were part of this inner circle. Stratton felt that Valley's study "shows a high level of technical competence in relatively narrow fields," but that air defense "involves economic and socio-logical factors quite as important as the purely technological ones and . . . no analysis to date has taken these properly into account."[36] He insisted that MIT undertake another study of air defense by a group whose "professional competence matches the standards established by Hartwell and Troy."[37]

Thus, in January 1951, Stratton and Killian agreed that the Air Force could sponsor the development of an experimental radar network around Whirlwind under the interim management of MIT's Research Laboratory of Electronics (RLE), but insisted that a longer-term project would only be established on the basis of a much broader study. They recruited Loomis to lead the new study group, comprised of twenty-eight scientists and engineers, in what was known as "Project Charles."[38]

1.4   Promising Programming Flexibility

The renewed study put the onus on Forrester to prove that a real-time digital computer could contribute to national security. Whirlwind was not a prestige project. As late as March 1951, Vannevar Bush advocated automating elements of air defense, seemingly unaware of MIT's work on the problem.[39]

Forrester's first goal was to prove that Whirlwind could coordinate the interception of Air Force airplanes and hostile bombers. A three-dimensional interception was out of the question. So his group prescribed the height of the "hostile" aircraft and programmed the computer to conduct its calculations in two dimensions. On April 25, 1951, the group used precisely timed flights to demonstrate three successful "interceptions."[40]

But Forrester needed to do more than demonstrate the feasibility of Whirlwind. He also needed to make the case that an electronic digital computer was the *best* way to coordinate a complex system. He argued that "a digital computer best meets the requirements for speed of operation and flexibility of set up to meet new and unforeseen developments," because of its "programming flexibility."[41] Rather than building complicated electronic equipment to connect each new piece of hardware, engineers would just write a computer program. Systems integration would become easy.

As an electrical engineer, Forrester was far more concerned about hardware than programming. He acknowledged that the advantages of digital computers—"high efficiency and flexibility"—came with "a vulnerability to electronic failure that is greater than in other systems."[42] Whirlwind's complex circuitry filled thousands of square feet in the Barta building at MIT and required frequent checking and tests. By August 1951, Whirlwind was operating reliably enough to be the basis for MIT's new Digital Computer Laboratory, but the group still anticipated random failures every five to ten hours.[43]

Forrester also had a strong interest in promoting "programming flexibility." But he might have known better. Although Whirlwind could be programmed to interact with equipment in real time, it could not be programmed in real time. Programming was as tedious for the Whirlwind as it was for any other computer. Following Wilkes's lead, they disbarred programmers from actually using the machine, requiring them to submit their programs to the Whirlwind operators, deduce program errors from printed results, and then wait a few hours or a day for new results.[44] In his concluding remarks at the first Joint Computing Conference of 1951, Forrester acknowledged that "time estimates in the digital computer field have been a standing joke," and that programming seemed to be the next big hurdle.[45]

But at MIT, Forrester's first priority was to demonstrate the promise of digital computing for air defense, and he succeeded. In August 1951, Project Charles's final report noted plans for a new, Air Force–operated air defense laboratory, soon known as Lincoln Laboratories (for its location 15 miles off-campus, in Lincoln, Massachusetts). It recommended developing computerized command and control centers to coordinate radars and antiaircraft weapons within each air defense division, and to communicate with other sectors across the continental United States.[46] In what was eventually dubbed the "Semi-Automatic Ground Environment," the battle would be coordinated by real-time computers at the heart of each command and control center.[47] If the Air Force gave the program sufficient priority, they anticipated that a "full-scale centralized system covering approximately the area of a present Air Division can be installed before the end of 1956."[48]

Top physicist-advisors had only a foggy notion of what they were getting into with digital computing. A newly formed Lincoln Steering Committee chaired by Loomis, and including elites like Zacharias, Wiesner, and Hill, tried to get a grip on the problem as they organized the lab in the fall of 1951. At a committee meeting in October, Loomis informed Forrester that powerful, higher-level advisors were requesting "the immediate construction of approximately ten Air Defense Computers."[49] Forrester replied that this was impossible. No computer capable of air defense had even been designed! Zacharias asked for more information, prompting Forrester to describe the Whirlwind and plans for a more advanced computer that might handle air defenses. He outlined a four-year plan that would employ as many as 100 individuals at its peak, though he prophetically acknowledged the "estimate . . . may be somewhat low."[50]

1.5   Selling Air Defense in Washington

Most members of the Lincoln Steering Committee had little basis for questioning Forrester's claims; they were focused on other problems. For example, as members of Project Charles, Zacharias led a section on passive defenses and Wiesner studied improvements to radar.[51] The gap between the advisory elite and the computer developers at MIT became an overt conflict as they organized Lincoln. Initially, Forrester directed the digital computer division, Valley headed a division on aircraft control and warning, and Hill directed work on communications and components.[52] In the spring of 1952, Loomis stepped down from his position as director, leaving Hill in charge. In April 1952, Valley objected to the arrangement and threatened to quit. He argued that most of the steering committee served "no necessary function" in Lincoln Labs, and only impeded "the freedom of action of the minority who are charged with the major responsibility . . . one which they should not have if it requires eleven other personages looking over their shoulders to keep them straight."[53]

Indeed, while Valley and Forrester scrambled to meet deadlines in the lab, Wiesner, Zacharias, and Hill were writing papers and flying to Washington, trying to shape national policy. The Air Force placed far more value on its Strategic Air Command than on its defensive systems, arguing that the best defense was a good offense. Zacharias, Hill, and other physicist-advisors recognized the deterrent value of offense, but also wanted protection for civilians. Furthermore, they felt that the Air Force was not planning ahead for the possibility of intercontinental ballistic missiles, and wanted to study the prospects of missile defense.

In the summer of 1952, Zacharias and other elite physicists tackled these problems in a Lincoln-managed study. They recommended deploying a long line of radars, extending hundreds of miles from Alaska to Greenland, to provide early warning of a Soviet air attack. They hoped that this Distant Early Warning (DEW) line would provide sufficient warning time for defenses to shoot down bombers, and for people in cities to evacuate or take shelter. The group emphasized the grim consequences of leaving cities undefended, outlining the attack that might materialize by 1954: a total of 100 bombs, dropped on urban targets in five different regions of the United States, would kill nearly twenty million Americans. Thus, the group recommended beginning immediate deployment so that the most northern portion could be completed no later than 1954, which came to be known as "the year of maximum danger."[54]

In pressing for early warning, the elite physicists went beyond the technologists at Lincoln, challenging Air Force priorities and goals. Furthermore, because they anticipated Air Force resistance, they took their case for early warning directly to the Truman administration. There they found an ally in Paul Nitze, the head of the Policy Planning Staff for Secretary of State Dean Acheson. A Harvard-trained economist, Nitze slowly shifted from Wall Street to Washington during World War II. Near the end of the war, he participated in the U.S. Strategic Bombing Survey, which underscored the value of early warning and air defense in the nuclear age.[55] In early 1950, Nitze helped draft a report for the National Security Council, NSC 68, advocating drastically escalating military spending to counter the threat of Soviet atomic weapons. As we will see, Nitze soon became one of the most persistent advocates of nuclear defenses.[56]

In October 1952, the National Security Council met with Lincoln Labs physicists, Air Force representatives, Paul Nitze, and others to discuss a continental early warning system. The physicists associated with the Lincoln Summer Study soon helped Nitze draft a National Security Council resolution recommending the urgent development of a distant early warning system. Though the Air Force continued to object, Nitze and Acheson persuaded Truman to commit the United States to a system that would provide three to six hours of warning "as a matter of high urgency" by the end of 1955.[57] This bureaucratic end-run led Zacharias, years later, to reflect that air defenses were "sold to Truman over the dead body of the Air Force."[58]

Selling it to Truman was not enough, for President Dwight Eisenhower inherited the controversy when he entered office in January 1953. Air defense was just one item that threatened Eisenhower's commitment to keeping a balanced budget, and ongoing studies did little to resolve his dilemma. The debate became increasingly public, with scientists like Oppenheimer, Killian, and Hill advocating for air defense in mainstream magazines and journals, while the Air Force sowed seeds of suspicion about a defense-minded conspiracy.[59]

What most settled the debate was the detonation of a thermonuclear device in the Soviet Union on August 12, 1953. In February 1954, the National Security Council emphasized that defenses, including "semiautomatic computers," were "vital to effective security."[60] Under a national security policy that came to be known as the "New Look," Eisenhower freed up funds for air defense by emphasizing nuclear-armed strategic bombers over conventional forces.[61]

2   Discovering the "Most Underestimated Task"

2.1   Systems Programming: Research or Production?
While the advisory elite was working to influence national policy, Forrester, Valley, and others at Lincoln Laboratories were working furiously on computerized air defense—and falling farther behind.

Competition only accelerated the hectic pace. In 1952, the University of Michigan's Willow Run research center proposed to develop an Air Defense Integrated System (ADIS) that would allow combat centers to exchange data electronically. Several Air Force officials felt that Michigan's project would be operational as an air defense system more rapidly than MIT's, and proposed funding the two projects in parallel.[62]

Forrester fought back. In November 1952 he prepared a revised plan for the Air Force, to correct "some mistaken ideas of the capabilities and delivery schedule" for his project. Now he called for an initial demonstration by September of 1953—less than one year away.[63] Forrester had learned to give "serious consideration" to such demonstrations, noting their "political importance."[64]

The stream of high-level visitors to the Whirlwind was plenty reminder of the urgency.[65] Yet September 1953 came, and went, without the planned demonstration. Though the technologists sometimes resented policy-oriented men like Zacharias, they were lucky to have friends in high places. Killian pressured the Air Force to drop negotiations with Michigan and put its full support behind MIT's air defense project.[66]

Nonetheless, the group continued to fall further behind. The complexity and concurrency of the air defense project made it tough for Forrester to see what was coming. The groups in his division worked to design, develop, and program three computers to integrate air defense technologies into three successive systems: the "Cape Cod System," the Experimental Sage Subsector (ESS), and the first Operational SAGE Subsector. Each computer and system grew more complex than its predecessor, yet there was no time to complete or test one system before beginning work on the next. In 1952, while Group 61 was still experimenting with Whirlwind I and the Cape Cod System, Group 62 was already designing the Whirlwind II (an IBM prototype, soon dubbed the XD-1) for the more complex Experimental SAGE Subsector. Tests of the XD-1 and the Experimental Subsector would not be completed before engineers began designing an even more complex production-line computer (the IBM FSQ-7) for an Operational SAGE Subsector.[67]

By 1953, the growing complexity of programming began to delay the Cape Cod System. The first demonstration of Whirlwind involved a single radar, interceptor, and target flying at a prescribed altitude. But programmers went on to link Whirlwind to multiple radars, and to conduct increasingly sophisticated tracking and interception tests. Since they were forced to share the Whirlwind with other groups, they vied for enough computer time to test their programs. Because the programs were tested in real time, in response to constantly changing conditions, it was not easy to replicate and fix problems.[68] As they confronted increasingly complex tasks, programmers struggled to write small and efficient routines for Whirlwind's small and unreliable computer memory, at times causing alarms "at the rate of four per hour."[69] This problem was only partly eased by the installation of the first-ever magnetic core memory.[70]

Yet schedules continued to slip. By November 1954, Forrester sent the Lincoln Steering Committee a 47-page memorandum requesting more "manpower" across the entire division. He estimated that integrating the XD-1 into Experimental SAGE Subsector alone would require about 27 programmers.[71] In early 1955, the group confronted the next step: how many programmers would be required to integrate the production-line FSQ-7 computer with all of the weapons for the first Operational SAGE Subsector? (See figure 1.1.)

Forrester's group was wary of taking on this programming job. They had typically focused on research, leaving "production engineering and final product design" of electronics to industrial laboratories.[72] However, they noted that the tasks of designing programs for a final, production-line system "are so interwoven with our responsibilities for systems guidance that no satisfactory dividing line for responsibility seems to exist."[73] Thus they took responsibility "for the technical content, as well as the doing" of master programs for the first SAGE direction center and combat center. Western Electric and Bell Labs, the two main systems contractors, would then adapt these master programs to individual sites around the nation.[74]

2.2 Programming "Manpower" and Slippage

In March 1955, Forrester and Everett reorganized and bolstered the group primarily responsible for preparing the master programs. John Jacobs, a graduate student in electrical engineering, was put in charge of a new section to coordinate the programming requirements with Bell Labs, Western Electric, and the Air Force.[75] Jacobs recalled that he quickly "became alarmed by the lack of planning on the programming."[76] Studying the problem more carefully, he projected that the master programs would require not 30 new programmers, but 200 new programmers. Still more would be needed to install SAGE subsectors around the country.[77]

This was a tall order in the mid-1950s. Most contemporaries estimated that there were only 200 skilled programmers in the United States. And most systems contractors for SAGE did not want to invest in training new programmers. As one IBM insider recalled: "We estimated that it could

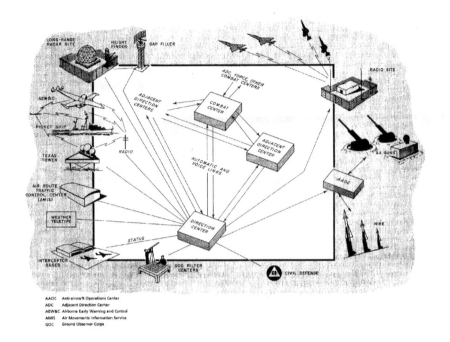

Figure 1.1
This diagram of the air defense system (circa 1955) illustrates the complexity of integrating weapons systems with the SAGE computers. The SAGE computers were located in combat and direction centers, and were responsible for coordinating radars, air defense weapons, and communications. Reprinted with permission of the MITRE Corporation.

grow to several thousand people before we were through. . . . We couldn't imagine where we could absorb 2000 programmers at IBM when this job would be over someday . . . "[78]

Jacobs finally turned to RAND, partly because it employed about 10 percent of the nation's systems programmers. In early April 1955, Jacobs and his division manager asked the head of RAND's Systems Training Program, M. O. Kappler, to take on the SAGE programming job. They agreed that RAND would send personnel to Lincoln Labs for training so that when Lincoln finished developing master programs for the first operational SAGE subsector, RAND could take over and adapt these programs to all additional SAGE subsectors around the nation. Kappler planned to grow a staff of about 70 people for the SAGE programming job, with additional programmers for systems training.[79]

Courses ran through the summer of 1955, and recruiting continued in earnest all fall. Training courses were shortened, and classes were overlapped, to gain workers more quickly. Jacobs recalled that "an 'experienced' programmer was someone who at one time had finished the act of writing a program."[80] Any programmer with this much experience might be charged with training new recruits. Personnel were taken away from ongoing work to "man the 'battle stations'" in recruiting and hiring offices.[81]

As David Kaiser has discussed, calls for more physics "manpower" were pervasive in the early Cold War.[82] Recruiting programmers was especially challenging because the work and skills of programming were so new. Herb Benington, a systems programmer for the Cape Cod system, recalled that it was "almost impossible to predict" good programmers before hiring.[83] Jacobs recounted "naively" thinking "that only engineers could become expert programmers," but discovering that individuals from quite different backgrounds, such as music and psychology, could often do just as well.[84] Women were very well represented in the burgeoning ranks of programmers (figure 1.2).

As Benington recalled:

. . . these people tend to be fastidious—they worry how all the details fit together while still keeping the big picture in mind. I don't want to sound sexist, but one of our strongest groups had 80 percent women in it; they were doing the right kind of thing. The mathematicians were needed for some of the more complex applications.[85]

In practice, there were many different kinds of work, and thus many kinds of workers, associated with programming. In a typical mid-1950s division of labor, top-level program designers would first carefully plan a sequence of instructions to be run by the computer. The program plans would then be handed off to more lowly coders—often women—who would translate them into machine-readable instructions, using keypunchers to represent binary code as a series of holes on cards or tapes. Although coders could use some equipment to ensure that the instructions were punched in as intended, neither programmers nor coders typically debugged programs in real time on the computer. Instead, they put the program cards or tapes into a queue, which computer operators—typically men—would run in a "batch." A few hours later, the coders or programmers would read the results of the run on a piece of paper (figure 1.3). If any errors appeared, the process would be repeated.[86]

This tedious process led to an acute shortage in programming "manpower," but it also opened up opportunities for women.[87] The same month

Figure 1.2
Madeleine Carey, a SAGE programmer, with a stack of 60,000 cards (the total number in the master program). Reprinted with permission of the MITRE Corporation.

Figure 1.3
Russell White of RAND, analyzing the results of a parameter program. Reprinted with permission of the MITRE Corporation.

that Jacobs began organizing the programming effort, managers in the digital computer project acknowledged a change in hiring policy. While they had previously tried to recruit individuals "of sufficient caliber to be potential section leaders," they now had "a definite need for some persons not of this caliber."[88] In January 1956, Group 67 was formed to develop programming utilities for the XD-1; it was staffed heavily by women.[89]

By the end of 1955, the programming problem seemed to be under control, with Jacobs' group anticipating "system operational checkout" of the Experimental SAGE Subsector (ESS) during the summer of 1956.[90] But no sooner did the master program seem within reach, than it flew out of control again, slipping from two, to three, to six weeks behind schedule.[91] In February 1956, Jacobs reported that "additional manpower will be needed to accelerate the effort since there seems to be no other way."[92] He couldn't hire more people because Forrester's group had already grown to the maximum allowed by MIT—534 personnel, twice what Forrester had proposed just a few years earlier. Asking RAND to hire programmers didn't resolve the issues since there was no place to put them. Programmers spilled into two hotels in Kingston, New York, eight large prefab buildings next door to Lincoln Labs, and a portion of the Army hospital next to the psychiatric ward.[93]

The group struggled on, through snowstorms, equipment failures, and fierce competition for computer time.[94] But managers seemed unable to get a grip on the slippage; just how far behind were they? It was not until the end of May 1956 that Jacobs delivered the bad news. The master computer programs, due just two months later, would be nearly a year late.[95]

Damage control began immediately. Forrester's group stressed that the delay was not a serious limitation, since equipment manufacturers were also behind schedule.[96] But, in fact, the delay forced a complete reevaluation of the SAGE system, and called its full deployment into question. The master program needed to be tested as part of the Experimental Subsector before it could be installed and tested in the first Operational Subsector—or any other subsector. Since the Air Defense Command never occupied the same priority as the Strategic Air Command, some in the Air Force viewed the programming slip as an opportunity to limit the scope and expense of SAGE. Soon a power struggle was brewing within the Air Force. By December 1956, General Partridge, commander of Continental Air Defense, was ready to go directly to the secretary of defense to express "as forcefully as I am able . . . my view that it is mandatory that we get the SAGE system running as soon as possible."[97]

2.3   Inventing a Whole New Art

Ultimately, those supporting a full and timely deployment of SAGE prevailed. Meanwhile, Forrester's group became more entangled with systems programming, prompting a professional identity crisis. As electrical engineers, they had imagined programming as something far more flexible than hardware. Jacobs later recalled: "There was a myth that you could do all or most of the integration of weapons by modifying the computer program. This turned out to be untrue. Not only did most of these changes require hardware changes, but the cost of programming soon rivaled the cost of doing the same thing with hardware."[98]

Indeed, at a Navy-sponsored symposium in June 1956, Benington put programming costs at $55 per instruction word. He later recalled that the "audience was somewhat chilled" by this news, since "the common goal was to produce instructions that cost less than $1 per line."[99]

Because the SAGE computer programs were unprecedentedly complex, Benington and his colleagues found themselves not with a simple engineering problem, but with a tremendous management job. Scaling up from a 35,000-instruction program to a 100,000-instruction program required hundreds of thousands of pages in documentation.[100] Everett recalled being "misled by the success we had had with capable engineers writing programs that were small enough for an individual to understand fully."[101] By contrast, SAGE confronted them "with programs that were too large for one person to grasp entirely and also with the need to hire and train large numbers of people to become programmers—after all, there were only a handful of trained programmers in the whole world. We were faced with organizing and managing a whole new art."[102]

While the SAGE engineers struggled to manage programmers, others were rapidly innovating in the new art. Most early programming innovations emerged from the cadres of men and women who developed and used digital computers for scientific calculations. For example, in 1952, Grace Murray Hopper, a mathematician who worked on computing machinery at Harvard during World War II, developed the first compiler—a computer program that could assemble complete routines from subroutines stored on a tape. John Backus, a mathematician at IBM, expanded on this technique when he developed the Formula Translating System (FORTRAN) for IBM's Defense Calculator. His system used a set of intuitive instructions (known today as a "higher order programming language") that would call subroutines from the library—and a compiler that would put together complete programs.[103]

Such "automatic programming" techniques could be controversial. Some programmers feared that their skills would become obsolete, and that they would be replaced by machines. However, Hopper suggested that automatic programming enabled professionalization.[104] And, indeed, new programming techniques did sometimes enhance the status of computer workers. For example, in the mid-1950s, several managers of IBM computer installations in the southern California defense industry formed SHARE, a group that aimed to save labor by developing more standard ways of programming, and exchanging commonly used routines.[105] As historian Atsushi Akera has discussed, SHARE gave programmers more autonomy and bargaining power with computer manufacturers and their industry employers.[106]

By 1958, the work and technologies of programming were so widely recognized that they earned a name: software. In the first known written use
of the term, the prominent mathematician John W. Tukey noted that "the
'software' comprising the carefully planned interpretive routines, compil-
ers, and other aspects of automatic programming are at least as important
to the modern electronic calculator as its 'hardware' of tubes, transistors,
wires, tapes and the like."[107] As historian Thomas Haigh has discussed, the
term *software* could refer to expert knowledge and practices as much as
to artifacts.[108] However, at the core of the new concept were techniques
for organizing complex processes, something that Stephen Johnson, in
historical work on Cold War systems engineering, has termed "procedural
knowledge." As Johnson notes, procedural knowledge was central to work
in fields such as project management, but was never seen as sufficiently
"fundamental" to give these fields academic legitimacy.[109]

Similarly, most scientists and engineers did not view software as a new
science or engineering field. Instead, programming was merely a set of
techniques for using computers, thereby enhancing the nation's technical
"manpower."[110] In fact, as historian Nathan Ensmenger has discussed, scientists, engineers, and business computer users were all uneasy with their
growing reliance on programmers. John Backus developed FORTRAN so
that scientists and engineers could communicate directly with machines,
eliminating the programmers altogether. He had little respect for programmers; years later he suggested that coders were transformed into programmers for "the same reason that janitors are now called 'custodians.'"[111]
When Backus reflected that programming was a "black art, a private arcane
matter," it was not a compliment.[112] Exceptionally good programmers were
unusually creative types—or so went the conventional wisdom in the 1950s
and 1960s, as managers struggled to find qualified hires. And because programmers were in perpetually short supply, they could gain special liberties
and relatively high salaries.[113]

3   Conclusion: Finding a Place for Programming

For the scientists and engineers charged with developing unwieldy code for
air defense, programming seemed more like "production" than research.
This did not trouble managers at RAND, who were rooted in the Southern California defense industry and could easily see an opportunity in the
SAGE programming job. By the end of 1956, they were planning a spin-off,
the Systems Development Corporation (SDC), which would be dedicated
to programming.[114] Between 1955 and 1957, SDC trained more than 700

programmers on the East Coast, while the Santa Monica programming group grew from 125 personnel in 1956 to 1,400 in 1962, with many more employees in the field. Programmers from SDC were recruited away by the higher-paying commercial sector nearly as rapidly as they could be trained; by 1958, employment termination rates were running about 20 percent per year. One manager recalled that SDC "trained the industry!"[115]

Initially, engineers at MIT were much more ambivalent than RAND about taking on software development; it smacked of production rather than research. All the same, they wanted to assure themselves of a future after SAGE, and recognized that they might be well-prepared to manage systems integration throughout the lifetime of SAGE, adapting the computer programs to include new weapons and technologies.[116] Indeed, in April 1956, the Air Force asked Lincoln Labs to take on the long-term work of systems integration.[117] By that time, Forrester was sick of managing administrative and personnel issues. In June 1956 he left Lincoln Labs to accept an MIT professorship in industrial dynamics, leaving his replacement, Everett, to respond to the Air Force's request. Writing to Valley, Everett insisted that there "should be no half measures."[118] Lincoln Labs should either take on the systems integration problem indefinitely, or refuse all future management responsibilities for SAGE, instead taking on research problems that the Lincoln engineers found to be interesting. "After much consideration," Everett concluded, "I feel we should get out and pay the price."[119]

MIT tried to get out, but the Air Force did not take no for an answer. Meanwhile, Everett and his group only grew more involved in programming and systems integration, and more convinced that it could not be divorced from research and development. In early June 1957, Everett proposed that Lincoln "undertake the task of systems engineering for the air defense ground environment . . . not for a limited time but indefinitely," and "undertake whatever expansion in size or budget may be required to perform the task successfully."[120]

However, authorities at MIT refused to take on systems integration, believing that it was too close to "the product end of development," and thus "inappropriate to management by an educational institution."[121] Eventually, they reached a compromise: MIT would spin off a new organization, MITRE, to manage systems integration. In 1959, about one-third of Lincoln Laboratories staff, including the majority of the digital computer group, gained independence from MIT as part of this new research and development group.[122]

Those charged with the systems integration for SAGE were pioneering the new field of systems programming. Once the Lincoln engineers finished the master programs for SAGE combat and direction centers, the systems programmers at SDC got busy adapting these to SAGE subsectors nationwide. In June 1958, the first Direction Center at McGuire Air Force Base reached "token readiness." But programmers encountered other challenges as they installed additional sectors over the following three years. The idiosyncrasies of each site required considerable adjustment. For example, the radar return program was originally written on the assumption that all the radars would be tilted upward. But one radar on a high mountain in the Detroit sector aimed slightly downward, and as a result "all hell broke loose in the program" when the system was first turned on.[123] The programs would be sent back to SDC in Santa Monica, but often fixes were slow, and frequent changes in operating procedures made for continual change.[124]

Although the nationwide installation took place at an average rate of six sectors per year, the incorporation became faster. The first center required about 100 programmers to install, the second one (at Stewart Air Force Base in Newburgh, New York) required 40 to install. After that about 15 programmers became standard for each installation, with about four remaining on site to maintain the programs.[125]

By the early 1960s, new organizations such as SDC and MITRE were growing skilled in developing and maintaining large complex software systems. The "flexible programming" of digital computers, a feature that Project Charles once predicted would make systems integration easy, had snowballed into something far larger. Engineers such as Forrester and Everett discovered that programming involved far more than specifying air defense systems requirements, even far more than managing a team of lowly programmers.

But most elite scientific advisors, men like Wiesner and Zacharias, had no interest in the new field of systems programming. They continued to view computers as tools for research and engineering in established disciplines, not as a new science. Nonetheless, as we will see in the next two chapters, their experience with SAGE cemented their commitment to computerizing nuclear operations. Scientific advisors recognized that there could be no other way to counter what soon became their central concern: the threat of intercontinental ballistic missiles.

2 Framing an "Appallingly Complex" System

One of the next things we have to try to do is to design a computer system, which is a big set of electronic hardware that does mathematics at a faster rate than it can be done in any other way. We have to design some kind of a system that will notice that some of these [missiles] are slowing down faster than others and automatically tell us that they are not the warhead . . . and therefore, shoot at this other one over here.

—Herbert F. York, congressional testimony, 1958

While Forrester's group wrestled with complex computer programs for air defense, Wiesner, Zacharias, and other elite advisors were studying a more daunting threat: intercontinental ballistic missiles (ICBMs). Bombers could carry nuclear weapons around the globe in about 10 hours; ICBMs would shorten that to just half an hour. Furthermore, scientists and engineers could easily imagine ways to fool any missile defense system. After a few short minutes launching into space, missiles could release decoys or lightweight chaff to hide warheads, making defense impossible during the 20-plus minutes required to travel around the globe. Warheads would then only become visible in their last few minutes of flight, as decoys and lightweight objects slowed down in the earth's atmosphere, leaving warheads to plummet toward their targets at thousands of miles per hour.

Was defense against such a weapon possible? Scientists and engineers had enough experience with air defense to imagine missile warfare, and they recognized that missile defenses were an unprecedented challenge. In 1959, Jerome Wiesner and other presidential science advisors warned that missile defense was "appallingly complex in concept and in required performance criteria. No system which can approach such performance requirements has yet been demonstrated in test or practice."[1] Not only did engineers lack experience with such complex defenses, but the future was riddled with uncertainties. What would a Soviet attack look like? What kinds of defensive technology could be developed? And whom or what would it protect? These questions loomed large in the 1950s. There were no obvious criteria for evaluating the best path forward. So how did scientific advisors make sense of the problem?

This chapter shows how scientists' disciplinary repertoire—physics and electrical engineering—shaped the way they directed the missile defense program in the 1950s. As historians have noted, this was a critical and formative period, one that established an advisory system and missile defense program that lasted for decades.[2] While previous accounts have noted advisors' conclusions—they opposed the Army's proposed deployment of defense—this chapter looks more closely at how scientists reached their conclusions, and reveals their influence on far more than questions of deployment. By choosing to focus on some questions, while ignoring others, scientists shaped missile defense research and development in consequential ways.

In the late 1950s, President Eisenhower worried that the services were too torn by rivalries to develop the defenses that the United States needed, so he empowered the scientists he most trusted to serve as presidential advisors and directors of the missile defense program. As these scientists wrestled with the parochial interests of the defense department, they sought disciplinary objectivity. Physical laws and data enabled scientists to formulate questions for further research: How would nuclear blasts in the upper atmosphere affect radar transmission? How might the offense try to fool radars with physical tricks, like decoys or metallic chaff? What would radars see as nuclear warheads sped into the atmosphere from outer space? They recommended experiments and engineering research to answer these questions.

By the late 1950s, their assessment was grim: conventional defenses would not protect civilians from a determined attack, something that might eventually include thousands of nuclear missiles. Nonetheless, Eisenhower's science advisors continued to advocate the highest possible level of research and development for missile defense.[3] They hoped that limited defenses could protect the United States from a more limited attack, and advocated such defenses as a means of arms control. They believed that policymakers might agree to limit nuclear weapons if they knew that they could intercept some extra missiles that the Soviet Union might sneak into its arsenal.

As scientist-advisors well knew, guarding against such a surprise placed heavy demands on computers and their complex programs. Computers were needed to help radars detect a nuclear attack, track many speeding objects, distinguish real warheads from decoys, and direct the launch and flight trajectories of missile interceptors. Computers would have to do most of this automatically, without humans, and with little or no room for error.

But without a disciplinary repertoire associated with computing, scientists did not place a high priority on computing research. Instead, they concerned themselves primarily with the physical information that would be used by computers, treating computers as "black boxes" to be developed by industry. At the same time, they advocated the computerization of tactical warning systems, strategic warning systems, and even arms control systems. As we will see, physicists' advocacy of defenses and computerized command and control systems carried long-term consequences.

1   Racing against a Time Fraught with Danger

1.1   "A Very Marginal Proposition"

Although Wiesner, Zacharias, and other scientist-advisors began studying missile defense during the 1952 Lincoln summer study, they began devoting more serious attention to the problem in 1954. That year, they were spurred to action by Trevor Gardner, the "technologically evangelical" Air Force assistant for research and development.[4] Gardner felt that the Air Force needed to speed up its ICBM development program, so he formed a scientific committee to study the problem, choosing members for their potential political influence. The preeminent John von Neumann served as chair of what became known as the Teapot committee, a group that included Wiesner and other members of the advisory elite at MIT. In February 1954, von Neumann's group reported that the Air Force could develop an ICBM much more rapidly if only it organized the work properly. Gardner forwarded the report to the Air Force, underscoring the "urgency" of developing ICBMs.[5]

Gardner also inspired the Science Advisory Committee of the President's Office of Defense Mobilization (ODM), a group formed in 1951, with what committee member Killian described as "a full charge of ideas about the inadequacy of our military technology."[6] In 1954, Killian began leading the "Technological Capabilities Panel," a study of ways that technology might lower the risks of surprise attack. With the help of a steering committee, he organized some 42 civilian scientists and engineers around three projects: offenses, defenses, and intelligence. By the fall of 1954, the scientists were flying around the country for field trips, briefings, and meetings—all told, 175 gatherings.[7]

Everyone agreed that it would be ideal to kill missiles in the first few minutes of flight, as they launched out of enemy territory. As Gardner explained, "the best defense" would be "a 30-caliber bullet through the [rocket fuel] tank just before it takes off."[8] But the United States would need space-based sensors and weapons to kill missiles in this early stage, and in 1954, before any satellite had ever been launched into space, such defenses were sketchy at best.

Instead, scientists and engineers focused on using ground-based radars and missile interceptors to detect, track, and kill ICBMs. Killian's group recommended developing and deploying very high-powered radars in the far north, where they could detect ICBMs relatively early in flight. However, tracking missiles and their warheads precisely as they traveled through space was far more challenging. During World War II, engineers at the Rad Lab and elsewhere had developed a wide range of methods for confusing air defense radars, such as signal jamming and dropping lightweight metallic chaff to hide bombers.[9] They expected to develop similar techniques with missiles. For example, once in space, a missile could release metallic decoys and chaff along with the warhead. The decoys and chaff would move at the same speed as the warhead, while reflecting confusing signals back to radars. These and many other countermeasures were so readily derived from well-known principles in physics and radar engineering, that they would eventually be published in the open literature, as we will see in future chapters. Although scientists would continue to refine knowledge of potential countermeasures through the 1950s, basic physical principles enabled them to sketch several possibilities by 1954.[10]

Thus, members of the Killian committee expected that warheads would only become visible in their final few minutes of flight, as they plummeted down through the earth's atmosphere, while lightweight decoys slowed or burned up. During these final few minutes, very fast, short-range rockets with nuclear warheads might destroy such missiles by detonating nuclear warheads nearby. But this would produce blast effects and fallout at relatively low altitudes. Without major investments in fallout and blast shelters, city defenses would be impossible.[11]

The essential problem was one of time—and this made computers critical to defense. An attacker could spend years planning the perfect strike, while the defender would have only minutes to respond, with almost no margin for error. As von Neumann explained, since "the attacker can pull surprises and has selection of what surprises he pulls," defense "looks exceedingly handicapped and difficult."[12] He felt that defense "is a very marginal proposition and chances are the counter-measures stay ahead of it . . . defense is . . . possible, but I would rather think that the chances are not good."[13]

In March 1955, Killian and his elite steering committee delivered the bad news to President Eisenhower and some 75 high level officials at the White House. They used a timeline comparing the military strength of the Soviet

Union and United States to explain that the United States would retain the upper hand for up to a decade, but that both superpowers would eventually have enough nuclear missiles to break through any defense the other could build. However, they did not lay down arms. Instead they warned that this time would be so "fraught with danger" that the United States should pursue "all promising technological developments" to delay it.[14]

1.2   Defense at What Cost?

But what exactly constituted a "promising technological development" in missile defense? The services each answered the question somewhat differently, and by the time Killian filed his committee's report, the Air Force and Army were vying for control of missile defense. From its inception, the Air Force developed fighter planes that could defend large areas far away from their targets, while the Army focused on defending local areas, first developing the Nike-Ajax antiaircraft missile (with a range of 25 miles), and then the Nike-Hercules missile (range of 75 miles). But in the mid-1950s, Army and Air Force weapons began to overlap in range and capability. Furthermore, the Air Force did not want to rely on Army missiles for last-minute defenses, so it proposed using the Navy's short-range Talos interceptor. For its part, the Army did not want to rely on Talos to defend its bases, nor did it want its missiles under Air Force control. In 1955 General Stanley Mickelson, commanding general of the Army Air Defense Command, informed the computer developers at Lincoln Labs that the "Army is interested in having information, not orders, from the SAGE System."[15]

In November 1956, Secretary of Defense Charles Wilson tried to curb competition in air defense by assigning the Army to "point defense" (defined as ranges of 100 miles or less), and the Air Force to "area defense" (defined as ranges beyond 200 miles). But the competition extended to missile defense the same month, when the Army announced plans for its long-range "anti-missile," the Nike-Zeus.[16]

The rivalry also opened questions about the relationship between air defense and missile defense. Would the electromagnetic radiation from the radars of missile defense interfere with the radars of air defense? Would air defense and missile defense share computational facilities? Jerome Wiesner led a committee in addressing these questions. In March 1957, his group acknowledged that the high powered radars required for missile defense posed "grave danger" of interfering not only with air defense radars, but also systems for air traffic control, television, and so on.[17] They also anticipated that local defenses would require relatively simple computers, but acknowledged that far more elaborate facilities, akin to the SAGE computer system, would be required to integrate air defense and missile defense into broad area defenses.

With the rivalry between the Army and the Air Force growing, Eisenhower's assistant secretary for defense research and development, C. C. Furnas, commissioned a review of missile defense efforts. Though the Air Force commissioned multiple conceptual studies of missile defense, which came to be known as Wizard, it made no specific development or deployment proposals, because "a good deal of quite basic research is yet required."[18] By contrast, a 1956 Bell Labs study showing that the northeastern quarter of the United States could be defended at a cost of $2 billion persuaded the Army to contract with the Labs for the Nike-Zeus interceptor.[19] However, Furnas's study group concluded that it would be "premature" to start developing missiles and local radars for deployment, tactfully surmising that the Army's proposed dates and cost "may be somewhat optimistic."[20] Furthermore, group members questioned "whether the effectiveness of the active system could be expected to be worth the cost in dollars, manpower, and resources and recommended that a careful evaluation of the advisability of doing it at all should be made."[21]

Indeed, the Eisenhower administration was beginning to question the ultimate value of defenses. By March 1957, the National Security Council's Planning Board noted that the United States no longer possessed "overwhelming nuclear superiority," increasing the chances of a Soviet attack.[22] The Defense Department expected that it would be difficult and costly to integrate missile defenses into the SAGE system. Even if air and missile defenses were highly effective, the United States had no means of preventing a multi-megaton bomb from being assembled at key places like the Soviet embassy in Washington, D.C. Nor could it prevent bombs from being smuggled into the port of a coastal city.[23]

Finally, new studies of nuclear weapons effects suggested that blast, radioactivity, and fallout from a Soviet nuclear bombing raid would kill between 53 and 91 million Americans. Anticipating that radioactive fallout would kill Americans far, far away from a nuclear detonation, the Federal Civil Defense Administration recommended spending $32.4 billion on shelters to protect civilians. This proposal "deeply troubled" Eisenhower's Planning Board, prompting four new studies of issues related to the problem, including a study by the Scientific Advisory Committee.[24]

1.3 "Offensive Airpower Is Always Successful"

The Science Advisory Committee was eager to get started on the new study, and Killian soon recommended his long-time colleague and friend Rowan

Gaither as director. Trained as a lawyer, Gaither had directed administrative issues at the Rad Lab during World War II, then became the chairman of the board at the fledgling RAND in 1948, where he began organizing a study of defenses in the summer of 1957.[25]

Although Killian and other members of SAC naturally saw a renewed study as a continuation of the Technological Capabilities Panel (the panel had recommended a follow-up study after two years), members of Eisenhower's administration did not want the civilian scientists meddling in military policy. In late June, Robert Cutler, Eisenhower's special assistant for national security affairs, and Donald Quarles, his deputy secretary of defense, directed Gaither to focus only on civil defense. Cutler wanted it to be "clearly understood that the Panel's mission does not extend to a detailed examination of national security policies and programs," and that he did not want a study like Killian's Technical Capabilities Panel. Similarly, Quarles told Gaither that his group should not "recommend the reorganization of a Continental Defense System which takes 10 years to bring about in the first place."[26] Cutler directed the panel to analyze three aspects of protecting civilian populations: active defenses, passive defenses, and "economic and political" rationales for investing in either or both kinds of defense.[27]

Formally, Gaither organized a study that hewed closely to Cutler's directive. Jerome Wiesner cochaired a panel on active defenses, while others studied passive defenses, and the socioeconomic and political implications of defense. But the goals of the study quickly expanded. One key influence came from systems analysts at RAND. Albert Wohlstetter, a logician and analyst at RAND began studying the vulnerability of the Strategic Air Command to surprise attack in 1952, and went on to campaign for more survivable ways of basing and fueling SAC bombers. In 1957, he helped persuade Gaither that civil defenses were less important than reducing the vulnerability of offenses forces.[28]

In the summer of 1957, Gaither requested permission to expand the study, arguing that economic considerations required a broader examination of military programs. Cutler later recalled giving "assent without foreseeing the result"—the group "was soon busying itself more about military than Civil Defense matters."[29]

As committee members began flying around the country for briefings in the summer and fall of 1957, the military managed the civilian scientists carefully. For example, when Wiesner and several other members of the steering committee traveled to the headquarters of the Strategic Air Command for briefings, they were treated to a display of the Air Force's offensive power. The chief of the Strategic Air Command, Thomas Powers, emphasized that Soviet air defenses could not significantly diminish the Strategic Air Command: "offensive airpower is always successful."[30] Similarly, the Air Force vice chief of staff, General Curtis LeMay, stressed that the main purpose of the U.S. air defense system was to provide detection and warning for SAC, since "there never has been nor is it likely that there will ever be a means of devising an absolute or completely effective defense."[31] Sprague and Wiesner were surprised by the briefings; they "had received an entirely different picture" during their visit to the U.S. Air Defense Command.[32]

By some accounts, LeMay had no intention of waiting for the six hours provided by tactical warning systems. There was a good reason: evidence suggested that SAC couldn't get off the ground in six hours. According to interviews conducted by nuclear historian Fred Kaplan, LeMay's secret solution was to act on strategic warnings gathered by unauthorized flights over the Soviet Union. LeMay reportedly told Sprague: "If I see that the Russians are amassing their planes for an attack . . . I'm going to knock the shit out of them before they take off the ground."[33] Apparently, Le May hoped that he would have a day or so of warning before the Russians could fully mobilize, giving SAC time to launch, get around the globe, and preempt the attack. When Sprague objected that this was "not national policy," LeMay replied: "I don't care . . . It's my policy. That's what I'm going to do."[34]

The electrical engineers and physicists who studied active defense felt that a better solution was to improve early warning radars. But they recognized that missile defense radars would be vulnerable to countermeasures. William E. Bradley, an electrical engineer and research director at Philco (one of the Lincoln contractors), directed a subcommittee on missile defenses. The group sketched a variety of technologies that could be sent along with warheads to confuse the radar. And even if intercontinental ballistic missiles could be countered, the committee saw submarine-launched ballistic missiles as a "formidable" threat "for which there is presently no known adequate countermeasure."[35]

1.4 Endless Race or Arms Control?

While physical analyses suggested the limits of missile defense, the most gruesome findings of the Gaither committee came from the systems analysts who were charged with calculating the lives that might be saved by shelters in the event of a Soviet first strike. The numbers showed that fallout shelters offered the most cost-effective protection from a nuclear attack: "no other type of defense is likely to save more lives for the same money."[36] Thus the panel on passive defenses recommended a nationwide, $25-billion fallout shelter program, noting that it would cut casualties nearly in half.

Nonetheless, massive death and destruction would be inevitable in a nuclear war. Even with the recommended shelter program, a Soviet attack could kill nearly 100 million people—well over half the U.S. population in 1957. With a $45-billion blast shelter program, casualties might be limited to 40 million people, or one quarter of the population. The report stressed that the United States had the upper hand in a nuclear war, but not for long. A timetable outlining the future threat anticipated a period in which "each side could deliver through the other's defenses" an attack that would "destroy approaching 100 percent of the urban population, even if in blast shelters, and a high percentage of the rural population unless it were protected by fallout and blast shelters."[37]

The Gaither study convinced Wiesner of several things: "it was not really feasible to protect the American people if a global nuclear war occurred," both sides "would suffer terribly," and "the job we had to do was to assure that such a war did not take place."[38]

None of this meant giving up on missile defense. As we will see shortly, Wiesner and his compatriots viewed defenses as a form of arms control. And arms control was a key element of the final report, which underscored "the great importance of a continuing attempt to arrive at a dependable agreement on the limitation of armaments and the strengthening of other measures for peace."[39]

But other members of the Gaither committee suggested that defenses were important as part of an arms race. Most notably, Paul Nitze, a long-time advocate of nuclear defenses, became first a consultant on, and then the primary author of, the final Gaither report.[40] The report began by pointing to "expansionist" Soviet intentions, backed by "formidable" military and economic capabilities, and made particularly threatening by the "dynamic development and exploitation" of technology.[41] It then recommended a set of measures that aimed to enhance deterrence and survival, including "a program to develop and install an area defense against ICBMs at the earliest possible date."[42]

The report emphasized that action should be not only *urgent*, but also *endless*. Killian's report had predicted that U.S. defenses would become ineffective once the Soviet Union amassed enough offensive weapons. By contrast, the Gaither report predicted that defenses would become decisive once both sides amassed enough weapons: a "temporary technical advance (such as a high-certainty missile defense against ballistic missiles) could give either nation the ability to come near to annihilating the other."[43] The report underscored the urgency of the technological race: "The missiles in turn will be made more sophisticated to avoid destruction, and there will be a continuing race between the offense and the defense. Neither side can afford to lag or fail to match the other's efforts. There will be no end to the technical moves or countermoves."[44]

By the time the various panels of the Gaither committee met to discuss their recommendations in late October, they could feel the winds of change. On October 4, 1957, the Soviet Union launched Sputnik, the first artificial satellite. Amid national clamor over a "technological Pearl Harbor," members of the Gaither committee had some cause to expect that Eisenhower might pursue the arms race more aggressively.

On November 7, Eisenhower heard the final recommendations of the Gaither committee. He heard about the gruesome consequences of nuclear war, and recommendations to negotiate an agreement to limit arms. And he heard recommendations for immediate and persistent action. The committee warned that within two years, the U.S. deterrent would be "completely vulnerable" to an attack by Soviet ICBMs: "If we fail to act at once, the risk, in our opinion, will be unacceptable."[45]

Wiesner recalled that at the end of the briefing, Eisenhower thanked the committee for its work, and reflected on his initial question: "If you make the assumption that there is going to be a nuclear war, what should I do?" Now Eisenhower felt that he had asked the wrong question. "You can't have this kind of war. There just aren't enough bulldozers to scrape the bodies off the streets."[46]

In the weeks after they briefed the president, members of the Gaither steering committee continued to discuss their possible impact. Wiesner told Gaither that he was "quite pleased" and attributed the report's acceptance "to the missionary work" carried out by Gaither and his associates.[47] But Nitze and others on the panel felt differently, like they'd failed to win converts. Tired of preaching to the unconverted, they leaked alarming conclusions. On December 20, the top story in the *Washington Post* began:

The still-top-secret Gaither Report portrays a United States in the gravest danger in its history. . . . It shows an America exposed to an almost immediate threat from the missile-bristling Soviet Union. It finds America's long-term prospect one of cataclysmic peril in the face of rocketing Soviet military might and of a powerful, growing Soviet economy and technology which will bring new political, propaganda, and psychological assaults on freedom all around the globe.[48]

The story was echoed in newspapers around the nation.

All of this turmoil contributed to what one member of the Gaither steering committee termed a "favorable environment"—that is, pressures for "an increased military budget."[49] And indeed, the launch of Sputnik, combined with the leak of the Gaither report, spurred the Army and the Air Force to escalate their rivalry over missile defense.

2 Toward Limited Defenses: Physics, Computing, and Bureaucratic Politics

2.1 Focusing Missile Defense Research: Physics and Radar

Just four days after the launch of Sputnik, outgoing secretary of defense Charles Wilson told Eisenhower that "trouble is rising between the Army and the Air Force" over the "anti-missile-missile."[50] A new secretary of defense, Neil McElroy, inherited this problem the very next day. McElroy decided to manage the rivalry by establishing a new agency to manage space research, including missile defense. On November 20, 1957, McElroy presented plans for what soon became known as the Advanced Research Projects Agency (ARPA), noting that it might "direct and manage" the anti-ballistic missile program.[51]

This effort to reign in the services backfired. The Army immediately went to the *New York Times* to make its case for a $6-billion crash effort to develop its "anti-missile missile." The Air Force contested the Army's promises in headlines the next day. The House of Representatives decided to investigate, opening special hearings in January 1958.[52]

The Army framed the missile defense challenge so as to highlight its key advantage: a working missile. It suggested that missile defense was a "duel" between a single missile and a single "anti-missile," emphasizing that its "missile killer" was "an actual development program in the hardware state—not a study program." Although the Nike-Zeus missile was only in a mock-up stage, and the earliest systems tests were not scheduled until 1960, the Army argued that a nationwide system could and should be deployed by 1961. It showed films of experimental interceptors destroying drone bombers. Although drones and jets would move much more slowly than would an ICBM, the Congressional committee was persuaded, and urged the secretary of defense to fund the Army's accelerated deployment.[53]

The Air Force soon reframed the problem of missile defense in a way that highlighted its own strengths in computing and radar. Air Force Deputy Chief of Staff Donald L. Putt pointed out that defenses relied upon "the computers, the communications systems, the data-correlation, the data-handling systems, the acquisition radars, and the tracking radars."[54] Noting that "the Army has never worked on an early warning radar," Putt argued that only the Air Force had already been working on the "total job" of missile defense.[55] The committee was equally impressed by the Air Force arguments, and ceased pushing for deployment of Zeus.

Eisenhower tried to manage interservice rivalries by empowering the civilian scientists whom he most trusted. Just two weeks after the launch of

Sputnik, he embraced the ODM Science Advisory Committee's suggestion that he establish a presidential science advisor and associated committee. In a special televised address on November 7, Eisenhower announced that James Killian would become the first presidential science advisor. Killian appointed most of ODM/SAC to the new Presidential Science Advisory Committee (PSAC).[56]

Eisenhower also put civilian scientists in charge of missile defense research and development. In 1958 ARPA lost the civilian space research mission to NASA, and instead came to focus on missile defense research. The Army was allowed to continue its developmental Nike-Zeus program, but ARPA took over basic research from all of the services. The following year, Eisenhower created the Directorate of Defense Research and Engineering (DDR&E), to coordinate research and development among all of the services, as well as ARPA.[57]

Missile defense research was strongly shaped by the man that Eisenhower chose to be ARPA's first chief scientist, and then the first DDR&E: Herbert F. York. When York went to Washington, he knew virtually nothing about missile defense. But he was a quick study. In 1952, just three years after finishing his PhD in physics at the University of California, Berkeley, he had become a founding director of Lawrence Livermore Labs. He went on to serve with more senior physicists on numerous advisory committees, and made an impression. Killian recalled that York possessed "a charming, boyish, and almost cherubic manner but a whiplash mind and courage."[58] York accepted an invitation to join the President's Science Advisory Committee shortly before his 36th birthday. Just a few weeks later, he became the first chief scientist of ARPA.[59]

York soon became a persuasive opponent of the Army's proposed missile defense deployment. In his first testimony before Congress, York countered the Army's emphasis on "anti-missiles," arguing that the "Nike-Zeus rocket is the least difficult part of the system . . . the Nike-Zeus system consists of radars, computers, other things of that sort, and it is on this side where the development is . . . not at the stage where it is ready to go ahead."[60] Similarly, York noted the need to design a computer to assist in discriminating warheads from decoys.

But while York and other physicists recognized that computers would be essential to defense, they did not advocate further computing research. Instead, in May 1958 York turned to Lincoln for help in radar and physical measurements research. Lincoln began to oversee what became known as the Pacific Range Electromagnetic Signature Studies (PRESS), which "piggybacked" on missile flight tests. By measuring and studying the electromagnetic spectrum produced by missiles as they streaked through the upper atmosphere at supersonic velocities, scientists could learn about prospects for detecting missiles and distinguishing them from decoys or other surrounding objects.[61]

York also sought the help of William Bradley, who had barely finished leading the Gaither committee's missile defense panel, when he began studying the flip side of the coin—how U.S. offensive missiles might penetrate Soviet defenses. Bradley's "Re-entry Body Identification Group" (RBIG) invented myriad ways of confusing radars, including booster fragments, chaff, radar blackout, and multiple reentry vehicles. Defense insiders noted that Bradley's group conducted "the most comprehensive and formal work" that had yet been done on missile defense, and credited him with establishing "almost all of the primary problems which would engage the technical community working on ballistic missile defense" for more than two decades.[62]

In the summer of 1958, two members of Bradley's group helped York sift through the many missile defense projects inherited from the services, choosing those that would most likely resolve the uncertainties associated with countermeasures. Initially, they carved up the Defender budget eleven ways, corresponding to the eleven staff overseeing the Defender program. One of these staff members was charged with overseeing data processing research, while the rest focused on physical issues such as the physics of the upper atmosphere, missile kill mechanisms, and radar measurements.[63] Recognizing that this was not necessarily the most logical division of research dollars, program directors continued to reorganize the program, in ways that prioritized physical research and development. By 1959, data processing research occupied just over 3 percent of the total Defender budget, while kill mechanisms and radar techniques occupied over 40 percent, and physical measurements, atmospheric physics, and discrimination occupied nearly 40 percent of the budget.[64] By 1960, data processing occupied less than 2 percent, while funding for basic physical research and radar development continued to grow.[65] After 1961, data processing research was not broken out as a separate item in the Defender budget.[66] And though data processing surfaced again and again in the ARPA directors' comments before Congress, it was typically lumped in with radar work.[67]

Indeed, ARPA's directors continually underscored their agency's work in radar. Perhaps most notably, ARPA continued the "Electronically-Steered Array Radar" (ESAR), a project it inherited from the Air Force. Engineers had long recognized that an array of small radio transmitters would produce a combined beam that could be steered electronically by introducing a phase shift across the array. In principle, such "phased arrays" would be capable of tracking many objects much more quickly than mechanically steered radars. Lincoln Labs engineers were skeptical about the viability of phased array radar, but the ARPA-sponsored project proved successful, and set a new paradigm for military and space systems.[68]

2.2   Nuclear Testing and Missile Defense

At the same time that York was helping organize missile defense research at ARPA, he was participating in PSAC's deliberations on missile defense. In 1958, the group pondered a complicated relationship between missile defense and nuclear testing. Atmospheric nuclear testing could address three outstanding questions about missile defenses. First, since an incoming missile would be destroyed by detonating a nuclear-tipped "anti-missile" nearby, engineers wanted to know about the effects of high-altitude nuclear explosions. Second, engineers feared that electronic fields produced by a nuclear explosion could effectively "blind" radars stationed on the ground, making the defense useless. And third, scientists were speculating that a nuclear explosion in the upper atmosphere might create an electronic "shield." If a sufficient number of electrons were trapped in the upper atmosphere, the control and guidance electronics of any ICBM attempting to leave or enter the earth's atmosphere would be destroyed.[69]

The only way to address these questions was to conduct high-altitude nuclear tests. But the United States had a decisive lead in nuclear weapons in 1958, and without a mutual test ban, the Soviets might catch up. Furthermore, based on what U.S. scientists could glean from Soviet atmospheric testing, it seemed that the Soviets' bomb design contained a flaw. The United States might exploit this flaw for defenses, but the Soviets would surely discover and fix it with further testing.[70]

Questions about nuclear testing came to a head in March 1958, when after three years of negotiation, and growing international protest about the hazards of testing, the Soviet Union declared that it would stop testing if Western nations did as well. Eisenhower turned to PSAC for advice. Would a mutual agreement to ban nuclear testing be in the best interests of the United States? And could Soviet compliance be verified?[71]

Recognizing the controversial nature of these questions, Killian shifted PSAC's monthly meeting to an Air Force Base in Puerto Rico, far away from Washington. There he asked the committee members whether they believed the nuclear test ban was in the best interests of the United States, and whether they believed it could be monitored adequately. In the first and only vote PSAC ever took, all members of the committee responded yes to both questions. All, that is, but the committee's youngest member, Herbert York.[72]

In 1958, York had barely resigned from directing the nuclear weapons laboratory at Livermore, and testing was core to the work and mission of weapons scientists. Furthermore, York felt that scientists should not be giving "political" advice. But after the meeting, Wiesner took him aside and "patiently explained" that "scientists could indeed do as well in handling this novel question as any other collection of experts."[73] York was persuaded, and the next day the vote became unanimous.

Upon returning to Washington, Killian told Eisenhower that the committee believed the United States should proceed with an impending test series, Operation Hardtack. This would provide information about radar blackout and the feasibility of an electronic umbrella defense. However, the committee concluded that the nation already possessed warheads that would work well for defensive missiles, and that the test ban would contribute to the "overall advantage" of the United States.[74]

Eisenhower overruled objections from his Joint Chiefs and members of the Defense Department and took Killian's advice. Three days after meeting with Killian, Eisenhower wrote to Khruschev and proposed an international "Conference of Experts" to study problems of verification. Khruschev soon accepted the proposal, and in the summer of 1958, delegates from around the world met in Geneva to discuss the challenges of nuclear test ban verification. The United States raced through its Operation Hardtack test series and then joined the Soviet Union in a nuclear moratorium in November 1958.[75]

2.3 Toward Limited Defenses

The Hardtack nuclear tests were disappointing; the number of electrons trapped in the upper atmosphere was too few and fleeting to create an effective "shield" against nuclear weapons. Additionally, PSAC's missile defense panel—chaired first by Bradley, and then by Wiesner—concluded that "the blackout problem . . . is even more serious than we had anticipated."[76]

Wiesner's panel also consulted with scientists and engineers who were crafting techniques to ensure that U.S. offenses could penetrate Soviet defenses, and who showed that decoys or chaff could easily confuse radars, while nuclear precursor blasts could completely blind them. Thus, in the fall of 1959, Wiesner's group warned that within five years, each incoming warhead could be hidden by anywhere from 5 to 30 decoys and "a large cloud of tank fragments."[77] They had already concluded that the Army's planned radars would be unable to cope with such countermeasures, calling efforts to develop higher frequency radars, which would be better able to see through a nuclear explosion, "totally inadequate."[78]

Wiesner's committee felt that the Army was rushing to deploy a defensive system that would be obsolete by the time it was completed. In the spring of 1959, the Army was preparing to test its Nike-Zeus antiballistic missile, using the Kwajalein Islands in the Pacific as a firing range.[79] With the new range barely approved, efforts to test the missile as part of a larger system of radars and computers could not begin until at least 1960. Nonetheless, the Army was already requesting funds to prepare for full-scale production of a missile defense system in 1961. PSAC urged the Army and its contractor to stop trying to meet "a production deadline," and to treat the Kwajalein test operation only as "a field experiment for AICBM research."[80]

In 1960, the Army renewed its request for funds to begin production of a Nike-Zeus system that would protect urban areas nationwide. Wiesner's AICBM panel rejected this proposal. As they had already warned Eisenhower: "We do not believe that an active anti-ballistic missile system can be made effective enough against a determined attack to provide significant protection for the civilian population."[81]

Nonetheless, the president's science advisors remained strongly supportive of missile defense research and development, as they sought to reorient the Army's proposals toward a different set of goals. They favored limited defenses as a form of arms control. Many wanted the United States and Soviet Union to negotiate a ban or limitations on ICBMs, and believed that limited defenses might facilitate such an agreement by mitigating the risks of cheating. As they explained, a limited defensive system might enable the United States to counter a few extra missiles that the Soviets secretly hid away, making "the consequences of limited clandestine production of missiles far less serious."[82] In both 1959 and 1960, PSAC recommended the highest possible levels of funding for research, and recommended that production funds be allocated as soon as the Army developed a plan for limited deployment of Nike-Zeus.[83]

Only one person seems to have resisted the deployment of limited missile defenses in the late 1950s and early 1960s: Herbert York. He viewed the Army's rush to deploy defenses as a product of interservice rivalries, a means of gaining a more central role in nuclear policies. And he feared that a limited deployment of Nike Zeus would only entrench the Army's bureaucratic interests, leading inevitably to pressures for a full-scale system.

York was all too familiar with the services' bureaucratic rivalries. For example, in 1960 the Army objected to using the Air Force's long-range ballistic missile as a target in tests of the Nike-Zeus system. The Army wanted to use its "Jupiter" intermediate range ballistic missile, partly out of fear that the Air Force might sabotage the test. As the Director of Defense Research

& Engineering, York refused to allow this change, arguing that the Jupiter was a poor substitute for a longer range missile. So the secretary of the army, General Wilber M. Brucker, called York before a group of Army brass and threatened him repeatedly. York felt a bit intimidated, but reminded himself: "I'm on leave from the University of California and there's nothing this poor so-and-so can do to me that I care about, and he knows it."[84]

Unable to budge York, the Army appealed to Secretary of Defense Thomas Gates. Gates referred the issue to George Kistiakowsky, a Harvard chemistry professor who succeeded Killian as the president's science advisor in April 1959. Kistiakowsky pulled together the "Ad Hoc Committee on Nike-Zeus," which concluded that the Jupiter was indeed a poor substitute for an ICBM.[85]

These sorts of pressures took a toll on York's health. One Sunday morning in August 1960, he had a heart attack.[86] During what became a six-week hospital stay, his successor as the director of Livermore Labs, Harold Brown, sent York a telegram: "Must you be precocious in everything you do?"[87]

Once York recovered, he resumed his opposition to a limited deployment of Nike-Zeus. In early November, Kistiakowsky mentioned that PSAC's missile defense panel favored "the very limited production of Zeus," and discovered that "York is violently opposed because he is convinced that once any production is started, it will not be possible to control it and we will have a fantastically large operation with very little return."[88] In his memoirs, Kistiakowsky speculated that "personal friendships" between scientists on Wiesner's panel and engineers at Bell Labs may have given "Nike Zeus the benefit of the doubt in the panel's report."[89]

2.4   Fail Safe?

One concern is conspicuously absent from all of PSAC's declassified discussions on missile defense: faulty computers. This is a striking omission. PSAC favored limited defenses that could intercept a few unexpected missiles, and recognized that with little time to respond, any defense against such a surprise would rely upon complex hardware-software systems. However, the group did not press for more research and development in computing; rather, they treated computers as something to be readily produced by industry. For example, when Wiesner led the PSAC panel on missile defense, he noted that the "system appears to have been well designed from a data processing point of view," and expressed "high regard for the competence of the technical staff developing the Nike Zeus equipment."[90] After this single vote of confidence, PSAC's missile defense studies did not mention computing again for over a decade.

Though Wiesner and other elite advisors did not press for computing research, they did advocate using computers in a wide range of command and control systems. In the spring of 1959, Wiesner led a PSAC study of the nation's rapidly emerging nuclear command and control system. The Air Force was in the process of constructing a Ballistic Missile Early Warning System (BMEWS), which would include three radars—in Thule (Greenland), Clear (Alaska), and Fylingdales (England)—linked by phone lines to the new North American Air Defense Command (NORAD) in Colorado Springs. Wiesner's group suggested several useful roles for computers in strategic and tactical warning systems, such as "sorting out satellites" and determining "which of a number of possible states of alert we should be in."[91] However, they did not record concerns about fallible computers.[92]

To be sure, defense experts recognized that command and control systems might fail, but emphasized the necessity of automation to cope with shortened times for decision making. For example, at RAND, Albert Wohlstetter acknowledged that accidents could result from "electronic or mechanical failures," the "aberrations" of humans in the chain of command, or "miscalculations on the part of governments as to enemy intent and the meaning of ambiguous signals." Wohlstetter's solution was to manage such risks of fallible systems through procedural reforms. He argued that since "unambiguous" evidence of a Soviet attack would not be available in time to launch the Strategic Air Command, the Strategic Air Command should be launched on less conclusive evidence of an attack, "with a fail-safe return procedure."[93] In this scenario, bombers would "return to base after reaching a predesignated point en route—unless they receive an order to continue."[94]

When the Air Force finally implemented the fail-safe procedure in March 1958, it raised some eyebrows. Members of the Eisenhower administration worried that the Soviet Union might discover the bombers en route and launch a retaliatory strike.[95] And indeed, the Soviet Union was quick to condemn the flights, noting that "the slightest mistake on the part of an American technician, from carelessness, miscalculation, or a faulty conclusion on the part of some American officer" could spark nuclear war.[96]

The Western press painted a rosier view. In April 1958, the president of United Press International announced the Air Force's "fail-safe" system by inviting the reader to imagine piloting a SAC mission that was launched on false information:

Do you proceed to your target, does your bombardier press the button and does the first nuclear bomb go "down the chimney" to start World War III? . . .

    No . . . You are saved, you and many others, by a powerfully simple plan called "Fail Safe." It is proof against error, human or mechanical.

Fail Safe, a term borrowed from the engineers, simply instructs you to proceed towards your target for a fixed number of nautical miles and then turn back if for any reason you do not at that point and at that moment receive coded orders to continue.[97]

The most controversial procedural change came in the Air Force's airborne alert operation. In 1959, Eisenhower authorized the Air Force to keep 12 weapons-ready bombers constantly airborne.[98]

As international tensions rose, many experts worried that fears of surprise attack might prompt an otherwise avoidable nuclear war. Thomas Schelling, a noted economist, nuclear strategist, and former member of the Truman administration, explained: "If surprise carries an advantage, it is worthwhile to avert it by striking first. Fear that the other may be about to strike in the mistaken belief that we are about to strike gives us a motive for striking." He worried about the results of "a modest temptation on each side to sneak in a first blow." [99]

Such concerns fueled an international conference on ways to prevent a surprise attack in November and December of 1958. Experts such as Wohlstetter and Wiesner joined a motley American delegation in Geneva. The surprise attack conference yielded little in the way of tangible agreements between the Soviet Union and United States, but it fueled debate about the most likely origins and consequences of a nuclear war.[100]

The anxieties of experts found popular expression. For example, in the book and then the movie *Failsafe*, a mechanical failure triggers a complex but unalterable sequence of events, ultimately resulting in the nuclear destruction of Moscow and New York.[101] Similarly, in *Dr. Strangelove*, a deranged Army general orders a nuclear attack on the Soviet Union, and all efforts to intervene in the command-control sequence fail.[102]

Whether catastrophe was initiated by man or machine, both movies called attention to risks within the automated, complex systems controlling nuclear weapons. Yet none of these movies called special attention to computer programs or software. Furthermore, most experts viewed automation as an inevitable requirement of war in the missile age. For example, when *Newsweek* reported that an Air Force B-47 bomber accidentally dropped an atomic bomb-in-progress on North Carolina, it also noted predictions that fallible humans would be supplanted by computers. An engineer at the Jet Propulsion Labs, William Pickering, explained:

The decision to destroy an enemy nation—and by inference our own—will be made by a radar set, a telephone circuit, and electronic computer. . . . Far more than being slaves to our machines, our very life depends on the accuracy and reliability of a computing machine in a far distant country. The failure of a handful of vacuum tubes and transistors could determine the fate of civilization.[103]

Similarly, the noted mathematician and founder of "cybernetics," Norbert Wiener, felt that it was "quite in the cards that learning machines will be used to program the pushing of the button in a new push button war."[104]

While the trend toward automation carried ominous overtones, most experts envisioned computers as tools to manage risk, not as sources of risk. In fact, after the Geneva conference, Wiesner drafted a report advocating "communication and data-processing facilities . . . for a disarmament control system," and suggested that "models of the control system" would be studied through a "computer simulation."[105]

3   Humans, Machines, and Nuclear Anxieties: Whither Software?

In sum: by the late 1950s, Wiesner, York, and other physicists had gained considerable influence over the direction of missile defense research, development, and deployment policies. As these advisors adjudicated interservice rivalries and bureaucratic parochialism, they drew upon physical data and knowledge, seeking a kind of disciplinary objectivity. Physics allowed scientists to calculate the trajectory of ballistic missiles, and to sketch out decoys and other ways of hiding nuclear warheads from radars. Using physical measurements and formal rules, scientists could estimate the effects of nuclear ionization on the transmission of radar signals, the kill radius of nuclear warheads, and the effectiveness of hardening retaliatory forces. Physical analyses demonstrated the vulnerabilities of the Army's radar, and suggested directions for further research and development.

As we have seen, physics was not the only factor that shaped scientists' advising on missile defenses. Scientist-advisors also considered the role that defenses might play in negotiations to limit nuclear weapons. These considerations led them to reorient the Army's defensive proposals toward a limited defense, something that might give the United States confidence to negotiate arms limitations. And when analyzed as *physical systems*, it seemed that missile defenses could protect against a limited attack, something that might come from an accidental launch, or small treaty-limited arsenals.

Had physicists been able to analyze defenses as *information systems*, subject to the challenges of systems programming, they might have seen problems even with limited defenses. Defenses could only ward off an accidental launch or small surprise attack by relying heavily on a complex computer program—a program that could not be fully tested before it was needed. But scientists did not express concern about the reliability of computer programs in the late 1950s.

Nonetheless, as the military began to pursue a proliferation of computerized systems, scientists and engineers began to recognize that software
development posed special challenges. Indeed, as we will see in the next
chapter, a diverse group of mathematicians, social scientists, and engineers
came to understand programming as a new kind of research. As a result, a
new "science" of software emerged from a place that the nation's scientific
elites least expected it: practical efforts to manage the growing complexity
of nuclear operations.

3   Complexity and the "Art or Evolving Science" of Software

No matter how complex a hardware system, except perhaps for NIKE-ZEUS [missile defense], a man is always there.
—Jack Ruina, testimony before the House of Representatives, May 1963

One congressman called the screen shots "the most interesting thing I have seen in 5 years." A warhead streaked from space into the earth's atmosphere, glowing brightly on the radar screen. Jack Ruina, the director of the Defense Department's Advanced Research Projects Agency, explained that by studying a full movie, scientists might find ways for missile defenses to distinguish real warheads from decoys. Another congressman was intrigued: "How long a time-span would that have covered?" Ruina paused to correct the misunderstanding. He explained that in a real nuclear attack "you would not have such a display." Time would be too short for human involvement: "a computer . . . makes all of the decisions."[1]

Ruina's comments that day focused on "Defender," ARPA's biggest research project. But they also touched upon a tiny new effort in command and control, soon known as the Information Processing Technologies Office (IPTO). Ruina, an electrical engineer and consultant to the President's Science Advisory Committee, explained that the new office would focus on the "'software' part of the computer sciences," rather than hardware, since "the great limitation" of computing "is really a concern of management . . . the difficulty of programming the computers."[2] From IPTO's inception he'd emphasized its focus on "fundamental questions related to the use of computers."[3] Now he elaborated: "How do you handle a computer better, or how do you match a computer with how the human brain works?"[4] Ruina explained that these questions would be answered by basic research into "the art or evolving science" of programming.[5]

Ruina's words suggest growing recognition of a "science" associated with computing. As we have seen, the physicists who helped launch SAGE in the early 1950s viewed programming as industrial production, not scientific research. But a decade later, talk of computer science was becoming more common. By 1958, Jerome Wiesner included computers as one aspect of a new science concerned with "the problems of organized complexity."[6] Like Ruina, he focused on the challenges of using computers: "If we wanted 10^8 or 10^9 bits of storage, IBM would gladly roll up its sleeves and provide it in a machine. The difficulty comes from the fact that we do not know how to use such capabilities . . . this is the theoretical challenge of the field."[7]

Wiesner, like many elite physicists, privileged theory. Yet as Atsushi Akera has shown, some of the most innovative computing research emerged from practical efforts to align different institutional interests in the 1950s and 1960s.[8] Arthur Norberg and Judy O'Neill have shown how ARPA's new command and control office helped institutionalize academic computer science during this same period.[9] This chapter builds on these accounts by more fully exploring the relationship between shifting nuclear policy and computer science in the early 1960s. How did physicist-advisors such as Ruina come to understand the emerging science of computing? And how did their conception of the field shape their advice on missile defense—systems that would depend critically upon computers?

As we will see, early computer science was shaped by the practical interests of three distinct constituencies: military officers concerned with managing nuclear operations, scientists and engineers trying to "get the numbers out," and researchers interested in "artificial intelligence." The first director of IPTO, J. C. R. Licklider, aligned these interests when he advocated improved human-machine interactions, or "man-machine symbiosis," as the key to managing complexity. Ruina, Wiesner, and other physicist-advisors were persuaded by this vision, and came to see computing as a science of human-machine interactions.

Furthermore, I argue these conceptions of computer science ironically divorced missile defense and software research. Although the semiautomatic air defense system fueled the art and science of software, physicists anticipated that missile defenses would allow little or no time for human intervention—they would be completely automatic. As a field concerned with improving human-computer interactions, software research seemed irrelevant to missile defenses. Instead, software research flourished as military commanders sought to enhance control of the complexity of nuclear operations—especially offensive forces.

1   Managing Complexity in the Cold War

1.1   Command and Control from SAGE to Counterforce

The SAGE consoles offered an appealing vision of control, and the Air Force soon began to transfer this paradigm transferred throughout a wide range of increasingly complex operations. In March 1958, a lieutenant colonel from the Strategic Air Command approached the Systems Development Corporation (SDC) with a request: "General LeMay has seen SAGE in action and wants something like it, so that SAC commanders will know at any time where their forces are deployed."[10]

LeMay had certainly not seen SAGE "in action." The first SAGE sector would not become operational for another three months, and was never used in battle. Nonetheless, the Air Force soon contracted for the development of the computerized Strategic Air Command Control System (SACCS).

The new systems contractors actively cultivated Air Force interest in computerization. For example, in September 1958, John Jacobs, the director of the SAGE programming effort at MITRE, visited Colorado Springs to learn about efforts to move the North American Air Defense Command to the heart of Cheyenne Mountain. He then organized a study by MITRE, which concluded that automation of the NORAD Combat Operations Center was "imperative in view of the need for fast reaction to a ballistic missile attack."[11]

The Air Force soon issued a contract for the NORAD command and control system (425L). This was just one of many "Big-L" command and control systems modeled after SAGE (Project 416L), including the Strategic Air Command and Control System (SACCS, 465L), NORAD (425L), a weather reporting system (433L), an intelligence system (438L), and the Ballistic Missile Early Warning System (BMEWS, 474L).[12]

Defense Department managers recognized that this proliferation of command and control systems—each with its own computer and language—might create more complexity than it resolved. In November 1959 the Air Force asked MITRE to organize a study of best practices. This "Winter Study" recommended hardening command and control systems so that they could survive a first strike and developing new computer systems in an evolutionary way—slowly modifying systems rather than overhauling or commissioning entire new systems.[13]

As we will see, recommendations for survivability and evolutionary development would be echoed for decades. Each presidential administration would try to distinguish its unique nuclear posture and command and control philosophy, but in practice each would be constrained by the tangled and imperfect technological systems that the services had evolved.

Indeed, in January 1961, just months after the Winter Study issued its report, President John F. Kennedy's secretary of defense, Robert McNamara, inherited a jumble of command and control programs. McNamara seemed ideal for managing such systems. Trained in economics and business administration at Harvard, McNamara helped establish an office in systems analysis, which improved the Air Force's bombing efficiency during World War II. He transformed management practices at Ford Motor Company after the war, and became the company president at the age of 44. Only a month later, Kennedy asked him to become the secretary of defense. McNamara soon accepted. Known for his "awesome . . . ability to stay abreast of facts and figures," McNamara intended to improve the Defense Department's efficiency.[14]

McNamara's job was especially difficult because he and Kennedy rejected Eisenhower's nuclear doctrine of "massive retaliation." They wanted weapons and plans that would allow them to respond less destructively to a Soviet first strike. By February 1961, McNamara had embraced the recommendations of RAND analysts who advocated a "counterforce" targeting doctrine, pitting weapons against weapons, rather than against cities. Counterforce placed heavy demands on command and control. In his first budget speech to the Congress in March 1961, President Kennedy called for command and control systems to become "more flexible, more selective, more deliberate, better protected, and under ultimate civilian authority at all times."[15]

Meanwhile, counterforce provided the services with a rationale for flooding the new administration with requests for new weapons. For help in evaluating these proposals, the new administration turned to established physicist-advisors. Jerome Wiesner became Kennedy's science advisor in February of 1961. Similarly, McNamara invited Eisenhower's director of defense research and engineering (DDR&E), Herbert York, to continue in this role. But after suffering a heart attack at age 38, York recognized that he could not continue to work under the pressures of Washington politics. He lowered the pressure just a bit by becoming the first chancellor of a new University of California campus in San Diego. Harold Brown, the physicist who in 1958 succeeded York as the second director of Lawrence Livermore Labs, also succeeded York as the second DDR&E in April 1961.[16]

While McNamara took advice from physicists like Wiesner, York, and Brown, he relied most heavily on systems analysts. He persuaded Charles Hitch, director of the economics division at the RAND Corporation and president of the Operations Research Society of America (ORSA), to become the Pentagon comptroller. Hitch invited economist Alain Enthoven, a former colleague at RAND, to head a new Office of Systems Analysis. The systems analysts came to be known as "McNamara's band," and the new secretary of defense got them started by blasting "96 trombones" on March 1, 1961, a list of major questions assigned to various offices in the Pentagon.[17]

McNamara's sixth "trombone"—on nuclear command and control—returned a worrisome review. In August 1961, McNamara noted "serious deficiencies" and commissioned a task force to recommend improvements in command and control.[18] Named for its director, the retired Air Force general Earle E. Partridge, this new study was strongly influenced by consultants from MITRE and RAND. While the Partridge Report reiterated long-standing recommendations to computerize military command and control systems, it also sparked controversy over a new concept: an integrated, unified system under civilian control, the National Military Command System (NMCS).[19]

The services disapproved of a command system that would give civilians an "over-the-shoulder capability," blaming failures like the Bay of Pigs on "the intrusion of high-policy groups into the operation itself."[20] Nonetheless, in June 1962, Robert McNamara directed the Joint Chiefs to work out the requirements of a civilian-controlled, computerized, and integrated command system.

In practice, the services retained considerable autonomy in setting their own systems requirements, leaving the NMCS relatively fragmented.[21] By 1964, *Armed Forces Management* cataloged 35 SAGE-like "L-systems" along with dozens of other military automation projects, many of them using incompatible computers.[22] The Air Force pursued new systems most aggressively. In 1964, *Air Force Magazine* promised "a new degree of automation that may permit man to remain the master of his complex affairs in the age of nuclear weapons, ICBMs, and the population and knowledge explosions."[23]

1.2   The Making of a "Management Monstrosity"

Although the SAGE control rooms became showcases for members of Congress and high officials in the defense department, close observers questioned the efficacy of the new systems.[24] In 1964, one researcher at General Electric noted that "many SAGE sites are not exercising their equipment from one inspection to the next . . . some have even put cellophane overlays over their scopes and returned to essentially manual operation." He asked: "What confidence can we put in the development of future command and control systems when SAGE is named as the great prototype?"[25]

The editors of *Air Force Magazine* insisted that "modern intelligence operations would be impossible without computers."[26] But Les Earnest, a MITRE engineer who helped develop the automated intelligence system

(438L), later recalled that the system was useless because the "the tasks of data entry, error checking, correction, and file updating took several days," leaving the computerized database perpetually out of date. By contrast, the manual system relied on "people reading reports and collecting data summaries on paper and grease pencil displays . . . and provided swift answers to questions because the Sergeant on duty usually had the answers to the most likely questions already in his head or at this finger-tips."[27] Earnest recalled that when senior Air Force staff members made "the embarrassing discovery" that nobody was using the automated system, they ordered officers to "ask at least two questions of the 438L system operators" during each shift.[28] Officers dutifully began to query the system twice per shift—but no more. After working for eleven years developing military systems, Earnest decided that he'd never seen an L-system that improved the manual system it replaced: "The implicit goal became 'to automate command and control,' which meant that these systems always 'succeeded,' even though they didn't work."[29] A former colleague agreed: "the emperor had no clothes."[30]

Nonetheless, computerized command and control systems were such potent symbols of high modernity that the business world soon sought to emulate them. In 1964, when *Fortune* ran a 10-month series on "The Boundless Age of the Computer," the SAGE system was featured along with its most prominent spin-off, an automated airline reservation system.[31] IBM began to market this "Semi-Automatic Business Research Environment" (SABRE) to American Airlines while still working on SAGE. In 1959, American Airlines wagered a staggering $30 million on the SABRE development contract, expecting that the automated reservation system would save millions of dollars per year.[32]

SABRE was soon over budget and behind schedule. Robert Head, a senior systems engineer, later attributed the problem to IBM's "flawed" marketing strategy, which "focused exclusively on equipment costs with no recognition of the enormous hidden costs in programming each device." Indeed, IBM rented facilities to house growing legions of programmers. In 1962, Head was working in one such facility in New York when he looked out the window and saw the exercise yard of a local prison. He realized the "difference between the prisoners and the Sabre programmers—the prisoners knew when they were going to get out!"[33]

When the nationwide SABRE system was finally declared fully operational in December 1965, its cost had far exceeded initial estimates. Furthermore, the first day SABRE went online, hundreds of passengers discovered that their virtually assigned reservations did not exist in reality.[34] Nonetheless, SABRE eventually yielded a significant return on the investment, and triggered an industry-wide effort to automate reservation systems.[35] And despite criticizing IBM's marketing strategy, Head remained confident of the value of automated systems, going on to found the Society for Management Information Systems in 1968.

As Thomas Haigh has discussed, "systems men" such as Head promised executives that the "total system" approach to management would provide them unprecedented corporate control—though "total systems" remained more rhetoric than reality.[36] Systems men emphasized that their knowledge was superior to that of programmers, but both challenged traditional occupational hierarchies and created tensions in the workplace.[37]

This tension was especially acute in the arena of command and control. In 1964, *Armed Forces Management* praised the operators of the NMCS for handling "more than 200 crises of one size and variety or another during the past year," but attributed the success "to sheer talent and energy figuring out ways to beat off its own alleged 'technical support.'" The magazine blamed the "outfits responsible for building" the system for "an organizational, procedural, and management monstrosity." It summarized the users' complaints:

1. There are far too many people in the act . . . and the man with the least power is the guy who has to make NMCS work.
2. The operator side of the house is out-manned, out-gunned and overwhelmed by the technical side of the house. This is, in turn, a grotesque waste since the technical boys can justify their existence only by building a big system. . . .

The article objected to "some 27 men in the Joint Command and Control Requirements Group, nearly 100 more in DCA [the Defense Communications Agency], backed up by 65 men under contract from Mitre Corp., all writing requirements papers on NMCS."[38]

By the mid-1960s, policymakers and managers acknowledged that command and control systems were "enormously expensive and relatively ineffective," but did not blame technical experts.[39] Instead, they critiqued the "weapons system approach" to developing computerized systems: "Under this philosophy, the commander developed his requirements and the technical agency put together a development plan. . . . From there on, it might as well have been a missile or an airplane."[40] McNamara stressed a "new" approach: "capabilities will be added piecemeal and the command system will be constantly in a state of evolution, and never quite complete."[41] As we have seen, the Winter Study had advocated evolutionary development years earlier; McNamara simply pressed commanders to oversee the process more closely. Meanwhile, with growing concern that "software is the greater of the two 'ware' problems," systems contractors like MITRE continued to have plenty of work.[42]

1.3 Improving Programming Efficiency

The growing scale of computerization fueled the rise of a systems programming industry, and with it, concerns about a shortage of programming "manpower." Programmers were needed not only for real-time control systems, but also to "get the numbers out" for scientific and engineering research. Efforts to improve the efficiency of programming drove at least two distinct research agendas: time-shared computing, and higher order programming languages.

Time-shared computing emerged as an alternative to the slow, inefficient programming process described in chapter 1.[43] Since machines were too expensive to dedicate to any single user, programmers would submit programs to a computation center, where they would be processed in batches. Programmers might wait for hours or days to receive results. In 1957, Robert Bemer, a programming researcher at IBM, described a different possibility. He envisioned an extremely large and fast "centralized computer with various lines radiating out" to terminal facilities, and likened it to a drive-through diner:

... it would be like a short-order cook. It takes the orders off the lines and, so to speak, heats up the griddle and sees that the toast is ready while it is pouring the coffee. It will be self-scheduling, self-regulating, and self-billing to the customer on the basis of use of the input-output device.[44]

Similarly, Walter Bauer, a researcher with TRW, proposed to help resolve the problem of "insufficient manpower" in programming by developing a computer utility.[45] He explained: ". . . each large metropolitan area would have one or more . . . super computers," capable of handling many tasks at the same time.[46] "Organizations would have input-output equipment installed on their own premises and would buy time on the computer much the same way that the average household buys power and water from utility companies."[47]

Other groups sought to improve efficiency by standardizing programming languages. For example, in the defense industry, computation centers that used the same computer model (and therefore the same language) formed collaborative user groups to share common routines. In 1957, three such groups asked the Association for Computing Machinery (ACM)— the world's first professional computing society—to study the possibility of developing a more general programming language that would travel between different kinds of machines.[48] Around the same time, a European collaboration asked the ACM to help develop a "universal" programming language. The ACM readily agreed, and soon began sponsoring the development of the "International Algebraic Language," which became ALGOL.[49]

The Defense Department also had a strong economic interest in standardizing programming languages used for business data processing. In 1959, a group of representatives from academic and industry computing centers proposed developing a standard business computing language for defense contractors. The director of the Defense Department's Data Systems Research Staff, Charles Philips, confessed that his staff members "were embarrassed" not to have originated such a proposal.[50] Philips recalled that the Defense Department owned about 1,046 computers, with direct "software" costs of about $35 million in early 1959, and these costs "grew to over $200 million in the next five years . . ."[51] With Philips's support, the Defense Department began sponsoring the development of a Common Business Language (COBOL) in 1960, and it soon became a widespread standard.[52]

The systems contractors working on command and control also developed standard programming languages to improve efficiency. At SDC, a systems programmer named Jules Schwartz proposed to develop a new higher-order language for SACCS that would enable it to communicate with all of the other L-systems then in development. By February 1959, the SDC programmers were borrowing from Algol as they developed "Jules' Own Version of the International Algebraic Language" (JOVIAL). The new language significantly improved programming efficiency. The SACCS software system was much larger and more complex than the SAGE system—its operating system alone required over a million instructions, quadruple the size of the SAGE system—yet the project required less than half the programming cost and labor. JOVIAL eventually became the standard command-control language for the Army and the Air Force, and was also used extensively by the Navy and NASA.[53]

As Nathan Ensmenger has argued, languages evolved to help manage the complexity of the programming process.[54] Managers of computing centers were interested in time-sharing for a similar reason. But the "science of organized complexity" emerged from a somewhat different direction. Some mathematicians and social scientists were interested in computers for more than just "getting the numbers out"—they also valued digital computers for their ability to manipulate symbols, and thus to mimic human reasoning. Significantly, mathematician Alan Turing described a "universal computing machine" in 1936 because he was interested in logical reasoning, not rote calculation.[55] Similarly, von Neumann described a programmable computer as a "thinking machine" with "neurons" and "organs." By the late 1950s, mathematicians and social scientists who viewed computers as symbol-processing devices were carving out a new field: "artificial intelligence."

1.4 Toward the Study of Organized Complexity

Two leaders in the new field, Allen Newell and Herbert Simon, took inspiration from air defense research. In 1950, Newell left doctoral work in mathematics with von Neumann at Princeton to work at RAND, where he joined some psychologists studying how military organizations responded to stress. Newell and his colleagues decided to simulate an air defense direction center in a new Systems Research Laboratory, which aimed to practice "organizational theory in miniature."[56] The SRL soon hired Simon, a professor at the Carnegie Institute of Technology, as a consultant. Simon earned his PhD in political science at the University of Chicago in 1943, while borrowing liberally from fields such as economics, servomechanisms theory, and psychology, and went to Carnegie Tech in 1949 as the head of a new Department of Industrial Administration.[57]

Simon recalled that programming at the SRL—such as routines that could simulate radar maps—helped him understand computers as "capable of processing symbols of any kind—numerical or not."[58] This key insight was well-understood by mathematicians like Claude Shannon who noted that von Neumann's game theory—a mathematical model of conflict and cooperation among rational decision-makers—might be used to program digital computers to play chess.[59] However, Simon and Newell felt that game theory failed to capture how humans operated in complex, real-world situations. For example, military officers operating under the highly uncertain, time-constrained environment of a nuclear war could not know enough, or reason fast enough, to be so perfectly "rational." In a 1953 paper for RAND, Simon noted that "because of the psychological limits of the organism (particularly with respect to computational and predictive ability), actual human rationality-striving can at best be an extremely crude and simplified approximation to the kind of global rationality that is implied . . . by game-theoretic models."[60] Simon's elaboration of this "bounded rationality" eventually won him a Nobel Prize in economics.[61]

In 1955, Newell and Simon developed "the logic theorist," a computer program that could play chess.[62] Ed Feigenbaum, then a graduate student at Carnegie Tech, recalled that Simon walked into class and announced: "Over the Christmas holiday, Al Newell and I invented a thinking machine."[63] In the summer of 1956, Simon and Newell presented the logic theorist at a conference on "artificial intelligence," organized by a mathematics professor at Dartmouth, John McCarthy. The notion of "artificial intelligence" was controversial—Simon and Newell preferred "complex information process-ing"—but the label stuck.[64] By 1958, Simon, Newell, and their colleagues argued that digital computers had "opened the possibility of research, into the nature of complex mechanisms per se."[65]

Artificial intelligence researchers and managers of computation centers shared an interest in time-sharing and language development, but for some-what different reasons. Whereas managers wanted time-shared computers to improve efficiency, artificial intelligence researchers wanted better access to computers. By 1960, if a researcher submitted a program to MIT's com-putation center, he or she would be kept waiting, on average, one and one-quarter days before receiving the results. Matters generally got worse when undergraduates swarmed to complete their problem sets.[66]

Thus, when John McCarthy moved from Dartmouth to MIT in the fall of 1957, he was immediately interested in transforming the institute's cen-tral computer, an IBM 704, into a time-shared machine. McCarthy and like-minded researchers pressed IBM to modify the system so that it could respond to external signals, or "interrupts," in real-time. They then hooked the machine up to a flexowriter (a typewriter with electronic outputs) so that a single programmer could interact directly with the computer. In 1959, McCarthy and a new electrical engineering professor at MIT, Herbert Teager, proposed a scheme that would allow multiple programmers to time-share the computer.[67] Teager's and McCarthy's interest in time-sharing focused their organization of a "Long Range Computation Study" at MIT. In 1960, the study group recommended that MIT obtain a large, centralized computer that multiple users from across the campus could access and use remotely.[68]

While MIT's administration supported the proposal in principle, they couldn't pay for it. With little external funding, Fernando Corbato—a leader in MIT's computation center who as a physics graduate student in the 1950s had used Whirlwind "very hard"—helped develop the Compat-ible Time Sharing System (CTSS).[69] The CTSS consisted of multiple key-boards hooked up to MIT's IBM 7090 computer, coordinated by a schedule of interrupts. By May 1962, the CTSS could serve a few users in real-time—an ambitious accomplishment, but not nearly enough for a campus-wide system.[70]

While artificial intelligence researchers at MIT focused on time-sharing, those at Carnegie Tech focused on language research. Alan Perlis, a profes-sor of mathematics and associate of Herbert Simon's at Carnegie Tech, led the Algol language development. Perlis later recalled at least nine different computer languages between Carnegie Tech, Case Institute of Technology, and the University of Michigan: "each new computer, and even each programming group, was spawning its own algebraic language or cherished dialect of an existing one."[71] Academic institutions supported ALGOL because it could enhance the efficiency of scientific computing. By contrast, artificial intelligence researchers developed languages because they provided a way to study symbolic systems. Perlis later recalled that language research such as that at Carnegie Tech provided a way "for universities that had not previously been involved in computer design" to become "nuclei for the development of computer science."[72] Programming languages provided a theoretical foundation for academic computer science: the algorithm.[73] By the late 1960s, Perlis, Simon, and Newell would argue in *Science* that "the phenomena surrounding computers" were so "novel and complex" that they demanded a new field of study, "computer science."[74]

Nonetheless, computer scientists struggled to fund their most ambitious projects in the late 1950s. It took a charismatic researcher to unite diverse interests in computerized complexity—for command and control, "getting the numbers out," and artificial intelligence—around a clear vision and wealthy research sponsor. That person was J. C. R. Licklider.[75]

2   Structuring the Science of Command and Control

2.1   A Symbiosis of Research Agendas

Licklider earned a PhD in psychology in 1942, and then went to work in Harvard's psychoacoustics laboratory, an effort to understand the effects of noise on speech recognition. After the war he remained at Harvard and was caught up in the "tremendous intellectual ferment" surrounding mathematician Norbert Wiener's cybernetics, a new field that drew diverse scientists into discussions of feedback and control.[76] Licklider grew increasingly interested in modeling human intelligence, and in 1950 he got an offer he couldn't refuse. Leo Beranek, who directed Harvard's psychoacoustics research during the war, was starting up a new laboratory at MIT with another physicist, Richard Bolt. They invited Licklider to join them and build up a psychology department. Licklider went down the river, and in 1951, when Project Charles "needed one psychologist and twenty physicists," Licklider was the one.[77] He was soon conducting research for Lincoln's Human Resources Research Laboratory, examining how human-machine interactions might be optimized for air defense.

At MIT, Licklider built up an experimental psychology lab that aimed to model human intelligence using analog circuitry. However, he converted to digital computing after a transformative experience with a real-time computer developed in the air defense project, the TX2. The designer of the TX2, Wes Clark, found Licklider tinkering in a dark room in the Lincoln Labs basement, and invited the psychologist to try programming the TX2.[78] Licklider later compared the experience of real-time programming to "sitting at the controls of a 707 jet aircraft after having been merely an airline passenger for years."[79] The experience encouraged Licklider to abandon neural networks and to embrace digital computers as an alternative approach to cybernetics.[80] Furthermore, it encouraged him to view computers as tools for enhancing, rather than mimicking, human intelligence.

Licklider's experience helped him understand the shortcomings of SAGE. As a consultant to the Air Force in 1958, he noted:

The numerous human operators have been brought into SAGE mainly to handle tasks that turned out not to be practicable for the computer. It is therefore too much a matter of men aiding the machine, and not enough a matter of true man–computer symbiosis, to give us a preview of the Air Force information-processing and control systems that we hope will exist in the future.[81]

Licklider soon expanded his ideas in a conference paper on "Man-Machine Symbiosis." He avoided "argument with (other) enthusiasts for artificial intelligence by conceding dominance in the distant future of cerebration to machines alone," because his interests were more immediate: "There will nevertheless be a fairly long interim during which the main intellectual advances will be made by men and computers working together in intimate association."[82] In this view, computers were less interesting for their thinking abilities, than for their ability to help humans think. Licklider explained that interactive computers would help military commanders who had only a few minutes for making critical decisions. He also appealed to the scientific community, promising that they would be able to "think in interaction with a computer in the same way that you think with a colleague."[83]

Licklider's paper next drew existing research agendas around his vision of symbiosis. For example, he advocated time-sharing systems, and noted that language research would overcome "the basic dissimilarity between human languages and computer languages."[84] Licklider made no new technological proposals—McCarthy recalled "no surprises."[85] But rhetorically, it was brilliant. It provided a fundamental goal for computer research, one that united diverse interests in computers. Human-computer symbiosis, he promised, would help humans manage the complexity of scientific work, military operations, and much more.

Although Licklider is recognized as a leading visionary, evidence suggests that his influence emerged slowly. For example, when computer experts conducted a study of non-numerical information processing in the summer of 1961, they initially neglected "Man-Machine Symbiosis." Confronting proposals for a new institute focused on non-numerical information processing, Kennedy's science advisor, Jerome Wiesner asked Herbert Teager (who had been chair of MIT's Long Range Computation Study) to lead an evaluation of the field. Teager was joined by Anthony Oettinger, a mathematics professor who worked on language translation at Harvard's Computation Laboratory, and John Griffith, an IBM scientist who was developing supercomputers for the nuclear weapons laboratories.[86]

With an initial subject list ranging over command and control, information retrieval, artificial intelligence, simulation, and pattern recognition, the study group struggled to define "non-numerical" techniques. Nonetheless, a draft report noted shortcomings in all of these areas and criticized the military's "proliferation of development contracts," objecting to "growing external pressures towards large-scale hardware procurement" based on "vague" notions of "intellectual processing."[87] They called most work in artificial intelligence "a blind attempt to apply present-day computers to prove theories and ideas of shallow conception."[88] They highlighted software challenges, noting "very little work" on languages for non-numerical processing, and an "extremely difficult problem in maintaining any effective communication between software workers."[89]

When Wiesner circulated this draft report to colleagues near the end of 1961, it generated a firestorm of controversy. Artificial intelligence researcher Marvin Minsky objected to "the fantastic assertion" that the previous few years had seen little progress in software for non-numerical processing.[90] He pointed to languages developed for artificial intelligence research: "Since this whole area is only about 4 years old, and activity has been growing, it is hard to see what they mean."[91] Minsky also objected to the committee's critique that artificial intelligence research had produced "interesting specific results, e.g. checkers can be played and a set of theorems proved, but little of general applicability." Minsky replied: "This may be cute, but it is also stupid . . . The best work in the artificial intelligence area has concentrated on these problems because here the problem definition is reasonably clear."[92] Similarly, John McCarthy declared: "I am prepared to defend Minsky's and my work on artificial intelligence against their vague charges if this is relevant to any important decision."[93] McCarthy had in mind funding for a time-shared computer at MIT.

Reviewers unanimously agreed on the need for more focus and communication among software workers, but disagreed about the proper means for

achieving it. While academics defended their work as appropriately fundamental, Douglas McIlroy, an industrial researcher at Bell Labs, objected to
the panel's "wistful" emphasis on theory, noting: "The recommendation
that we concentrate on smaller, more tractable problems is nice, but in
an engineering situation (program writing has indeed become a large scale
engineering effort) such self-limitation is often unfeasible."[94]

Perhaps the most influential suggestion came from Herman Goldstine at
IBM, who pointed out that the study neglected work on the "man-machine
relationship."[95] Significantly, the revised report placed human-computer
interactions at the center of its analysis. It defined non-numerical processing more clearly by emphasizing analogies between humans and machines.
Whereas numerical calculations used "models of relatively well-formulated
and well-understood physical phenomena," non-numerical techniques
involved "the manipulation of symbols, meanings, and decisions."[96]
While those interested in numerical processing tended to treat computers
as "high-speed desk calculators," those interested in non-numerical processing tended to regard computers as "thinking machines."[97] The panel
criticized the notion of "thinking machines" for encouraging "costly large-
scale products of wishful thinking," but highlighted a middle road:

The most extreme form of the "desk calculator" view would not only deny the very
real assistance to scientific progress that computers have given to date, but it would
also close to exploration large and virtually untapped areas . . . such as . . . those
based on man-machine cooperation. . . .[98]

The final report recommended not establishing an Institute for Non-
Numerical Studies, and instead "expanding present programs in the computer sciences in universities."[99] Yet the group's restructured report, and
its positive reception, suggests that Licklider's notion of computer science
appealed to both academics and advocates of a new institute.[100] And indeed,
academic computer science began to expand rapidly after ARPA's new command and control office was placed under Licklider's direction.

2.2 Establishing a Science of Human-Computer Interactions

Licklider did not initially seek to direct ARPA's command and control office.
On the contrary, physicists managing Defense Department research sought
his help. They were receiving a preponderance of proposals for command
and control research, and the largest came from SDC, which was looking
for work as the SAGE project wound down. When the Air Force canceled
plans for Super Combat Centers for SAGE in 1960, SDC lost a major contract and four IBM Q-32 computers—developed especially for the Combat

Centers—became expensive "white elephants." So SDC asked the Defense Department for funding to buy a Q-32 and use it for a command research laboratory. The Q-32 was delivered in June 1961, and the same month, the Defense Department established what came to be known as the Information Processing Technologies Office (IPTO).[101]

ARPA's new director, Jack Ruina, didn't know what to do with this new office. A professor of electrical engineering on leave from the University of Illinois, Ruina was a radar expert and came to the job after helping Herbert York with a tricky aspect of BMEWS.[102] Ruina was familiar with ARPA's primary mission—missile defense—through his work on Bradley's Reentry Body Identification Group (RBIG). He knew relatively little about computers, but after a quick survey, he concluded that most defense department proposals for computer research were so "asinine," "such obvious baloney," that he was "turned off to the whole thing."[103]

Ruina felt differently after talking to Licklider. Interactive computing would clearly benefit physical research, while also improving military command and control. Licklider was not initially interested in Ruina's invitation to direct IPTO. Licklider had moved to the company started by his mentors, Bolt, Beranek and Newman, and was leading research on interactive computing. But he began envisioning the projects that might be supported on the Pentagon's dime, and talked himself into the job.[104]

Once Licklider assumed directorship of IPTO in October 1962, he suddenly enjoyed almost complete freedom to nurture his vision of human-machine symbiosis. He recalled: "Jack seemed too busy, he was just relieved to get somebody to run the office . . . [and] pretty much let me do what I wanted to do."[105] He recognized a bit of bait-and-switch: "The guys in the secretary's office started off thinking that I was running the Command and Control Office, but every time I possibly could I got them to say interactive computing."[106]

Licklider had good timing, for the Cuban missile crisis in October 1962 fueled concerns about command and control. On the day before Thanksgiving of 1962, Licklider shared a train ride to Washington, D.C., with Robert Fano, a mathematics professor who worked on MIT's long-range computation study, and who consulted for the military. Fano recalled bemoaning the "lousy" state of command and control with Licklider, concluding: "Something had to be done for the good of the country."[107] Furthermore, Licklider's enthusiasm for time-sharing rubbed off on him: "I got to feel that, my God, MIT should do something very important for the future. . . ."[108] Unfortunately, "nothing was happening" with MIT's time-sharing project.[109] John McCarthy had become so impatient that he went to Stanford.

With Licklider's inspiration and potential funding in view, Fano soon met with MIT president Jay Stratton and other administrators. To his surprise, they readily agreed to allow Fano to run an IPTO-sponsored time-sharing project at MIT—provided there was a space for the computer. Luckily, Fano knew of a space opening up, and by 1963 he was helping direct an IPTO-sponsored project known as Project MAC—alternately taken to mean "Multiple Access Computer," "Machine Aided Cognition, or "Man and Computer."[110]

Licklider continued to nurture research into interactive computing by selectively awarding grants around the nation. In 1963, he offered Simon, Newell, and Perlis $300,000 for research into computer languages. This move startled the researchers at Carnegie Tech; Newell had never even heard of Licklider.[111] Though Licklider recognized that many of the researchers he funded were more interested in artificial intelligence than "symbiosis," he funded whatever projects he felt would contribute to interactive computing.[112]

Licklider also deliberately limited funding for new machines, instead requiring a few centers to develop time-shared systems that could be used by multiple institutions. In November 1963, he proposed establishing Centers of Excellence in the Information Sciences, starting with MIT and Carnegie Tech. He reported that MIT was engaged in a practical effort, while the work on languages at Carnegie Tech would "develop the theoretical bases of information processing."[113] His high hopes were reflected in his label for the group of researchers he gathered to exchange ideas: the "Intergalactic Network."[114]

2.3   Divorcing Computing and Missile Defense Research

Licklider's work at IPTO helped establish academic computer science. All of the universities that established computer science departments in the 1960s received significant support from IPTO, and the three centers that Licklider privileged—MIT, Stanford, and Carnegie Mellon (previously Carnegie Tech)—were long-term leaders in the field.[115] By defining the new field as a science of human-computer interactions, Licklider gained academic legitimacy and military support. Physical scientists and engineers such as Ruina endorsed Licklider's vision because they recognized that it could enhance scientific "manpower" by making programming more efficient. At the same time, the science of human-machine interactions gained Defense Department support because it could improve command and control.

Yet by establishing computer science as a field concerned with human-machine interactions, Defense Department managers ironically divorced software research from the successor to air defense—missile defense. In congressional testimony, Ruina justified IPTO as important to virtually every command and control effort, *except for missile defense*: "No matter how complex a hardware system, *except perhaps for NIKE-ZEUS*, a man is always there."[116] As we have seen, Ruina envisioned a defensive battle that would involve no humans. He recalled that if "it was really a surprise attack," nuclear-tipped interceptors would fire without presidential authorization: "it was all up to computers."[117] Over forty years later, when asked if ARPA's software research was relevant to missile defense, Ruina replied: "Not at all . . . What drove that program . . . was the fact that the Air Force was going bananas and thinking of computers for all kinds of crazy ideas . . . our program tried to bring some sense into that business."[118]

Ruina also tried to bring some sense into the "Defender" missile defense program, viewing physical science and engineering as a corrective to the military bureaucracy. Ruina recalled: "The Army had interest only in developing a system for procurement. . . . The Air Force was still in its Buck Rogers period, and the stuff they were pushing in ballistic missile defense was absolute garbage."[119] He especially opposed "stupid" and "looney" proposals for space-based missile defenses.[120]

Ruina's deputy director, physical chemist George Rathjens, similarly felt that Defender had gone awry. Rathjens felt that Defender had become "a godawful mess," with "crazy work going on. . . . "[121] Most notoriously, in January 1959 ARPA launched its Guide-Lines Identification Parameters Program (GLIPAR), by inviting contractors to submit proposals for radical missile defense concepts, *assuming feasibility*. This program produced studies of exotic concepts involving antigravitation, antimatter, magnetic barriers, and "death-rays," drawing wide media attention and critique from the President's Science Advisory Committee.[122]

Ruina, a frequent consultant to PSAC, phased out studies of exotic defensive concepts. Ruina made physical measurements of reentry phenomena central, occupying 60 percent or more of the Defender budget. Radar projects also fared particularly well. Significantly, the reformed program did not include computing research. In fact, once Ruina reorganized and simplified Defender, data processing no longer occupied any explicit portion of its budget.[123]

Scientific advisors also drew on physics and bureaucratic savvy—but not the "science" of software—as they shaped Kennedy's policies on missile defense. With Ruina's help, Harold Brown assessed the missile defense program for McNamara, reporting in April 1961 that despite having invested $2.4 billion in defensive research and development, the prospect of "a really effective active defense of our urban population against a mass attack from the USSR is bleak. . . ."[124] Nonetheless, Brown followed the lead of PSAC's 1960 report by endorsing more research and development and emphasizing the value of limited defenses. He emphasized that partial defenses might provide "the environment in which an arms control agreement can become more acceptable due to its capability against small numbers of clandestine weapons."[125]

While scientist-advisors continued to advocate defenses to protect against surprise—a goal that would require completely automated systems in the final stages of attack—it seems that they took computer reliability for granted. A survey of all declassified reports in the 1950s and early 1960s shows that they never expressed concern about reliability. The nascent science of computing was concerned not with the reliability of code or machines, but rather with managing complexity through improved human-machine interactions. Thus, computer science was effectively divorced from research and advising on missile defense in the early 1960s.

2.4   On the Verge of Deployment: Defense as "Bold Play"

While physicists made the Kennedy administration well aware of the limitations of defenses, they also pointed to potential advantages for arms control, and these advantages were paramount in McNamara's reasoning during his first years in office. In 1961, hostilities grew over the Bay of Pigs fiasco. In August the Soviet Union resumed atmospheric testing, contributing to pressures for the United States to follow suit.[126] By the end of September 1961, McNamara was prepared to deploy the Army's proposed system, which had been downsized from a nationwide system of 70 batteries, to a six-city deployment of 12 Zeus batteries. McNamara allowed that the deployment would be too ineffective to justify by a "purely technical appraisal."[127] Nonetheless, physicists had also noted that the defenses might complicate a Soviet nuclear attack, defend against a small launch, and contribute to arms control, and McNamara found these advantages persuasive.[128]

PSAC continued to underscore the value of partial defenses, but began to fear that the Army had no intention of stopping with such a limited deployment. In October 1961, the new chair of PSAC's missile defense panel, Wolfgang ("Pief") Panofsky, sent "provocative" conclusions to Wiesner and McNamara: "neither the size of the proposed plan . . . nor its slow time scale (6 years) is supported by the stated reasons for partial deployment." Instead, Panofsky's panel suggested that "the size of the plan as presented is simply related to an arbitrary level of funding."[129]

McNamara continued to feel the Army's "tremendous heat" for deployment.[130] Thus, Wiesner arranged a meeting with Kennedy, Brown, and Ruina, but not McNamara. Ruina recalled getting "2 days' notice to meet with the president," on the day before Thanksgiving 1961.[131] There, Ruina showed Kennedy three hastily sketched alternatives to the Army's proposed system, arguing that Nike-Zeus should be improved by new technology from ARPA. Ruina described "Nike-Zeus-1" as a system with faster missile interceptors. These would enable the interceptors to destroy warheads closer to the ground, where they could be more easily distinguished from decoys. "Nike-Zeus-2" would include phased array radars, which could track many objects very quickly. By introducing a small phase shift across an array of small radar transmitters, these devices could steer their beams electronically rather than mechanically. He then outlined "Nike-X," a system that included phased array radars, faster interceptors, and other improved components.[132]

The meeting ran longer than expected. With a storm approaching, Kennedy was forced to end early, so that he could fly to Hyannis for Thanksgiving. But he invited Ruina and Brown to Hyannis, where they continued the conversation as part of a long series of meetings between Kennedy and his advisors. Ruina recalled that discussion of the defensive system was "a very short shift." When the subject came up, "The President said, 'Look, I've heard about this, I don't think we should go ahead with this system. Let's build that new system and forget the old one.' And he turned to McNamara and said, 'What do you think, Mac?' McNamara said, 'Sure, that's okay.'"[133]

Though Ruina felt that Nike-Zeus was abandoned in that one conversation, matters were much less settled.[134] Ongoing international tensions fueled Kennedy's desire to deploy defenses sooner rather than later. Unable to restart negotiations on a nuclear test ban, and with overwhelming public support for resumption of testing, Kennedy announced that the United States would resume nuclear testing in April 1962. Over the next eight months, the United States conducted a series of 36 nuclear tests. Meanwhile, missile defense entered the Cold War theater. In October 1961, Rodion Malinovsky, the Soviet defense minister, bragged that "the problem of destroying missiles in flight has been successfully solved," prompting applause from the Communist Congress and front-page headlines in the United States.[135] Two months later, the Army's Nike-Zeus successfully "intercepted" an antiaircraft missile, prompting more front-page headlines.[136]

As the saber-rattling continued, the Army underscored the urgency of a near-term defensive deployment.[137] In August 1962, the Army proposed deploying an "intermediate" defensive system with initial operational capability by 1967, and upgrading later. By contrast, waiting for the more advanced hardware recommended by Ruina would delay the earliest operational capability by two years.[138]

Some Defense Department advisors recommended accepting the Army's proposal. In early October, Daniel Dustin, a Lincoln Laboratories engineer and consultant to PSAC's panel on Anti-Intercontinental Ballistic Missiles (AICBM), acknowledged that Zeus could be overwhelmed by the Soviet Union's powerful nuclear offenses, and questioned whether "a serious nuclear-counter nuclear battle can ever be fought successfully with electronic sensors and control systems, however much they are 'hardened.'" He continued:

I would be upholding my standards as an engineer only if I said, "AICBM and CD [civil defense] make no sense."

. . . But it really troubles me to see that long gap of years with the U.S. having nothing to show in AICBM.

I finally conclude, laboriously and with reluctance, that some deployment of good old Zeus should be started. I would deploy it at a few of the most obvious urban centers with a bold play that it will defend them. Play is the right word, because the whole AICBM business is a game and will likely stay that way.[139]

Like most members of PSAC, Dustin used more than physics to evaluate defenses—he also drew upon his understanding of the timescales required to move from research to deployment, and the symbolic value of new weapons. Yet it seems that these physical scientists and engineers never expressed concern about software reliability, or the ways that slow, laborious programming might delay defensive deployments—issues that would come to the fore in the late 1960s. And in the early 1960s, the symbolic value of defenses nearly tipped the balance in favor of deployment.

3 Conclusion

PSAC's concerns about the bureaucratic momentum of defense won out in the fall of 1962. Despite strong objections from the Army, McNamara decided to replace Nike-Zeus with the longer-term Nike-X program.[140] A small group of senators attempted to speed up deployment in the spring of 1963, but failed.[141]

For the nation's scientific elite, the deferred deployment represented a triumph of scientific objectivity over the Army's vested interest in production. Ruina recalled that ARPA's Defender program kept "a small community of people involved other than Bell Labs in ballistic missile defense problems, and it is that community which defined Nike-X. Not the NIKE-ZEUS people. Those people resisted every step of the way. . . . They just wanted to build the goddamned thing. . . ."[142] Similarly, Defender director and physical chemist Charles Herzfeld argued that he helped correct a "crazy and irresponsible" program: "We based our stuff on good measurements and good theory . . . full size missiles and full size radars and all that."[143]

Yet, at the same time, elite advisors came to define the evolving science of software in a way that divorced it from missile defense research. As a science of human-computer interactions, software would enable humans to better manage complexity—whether of nuclear operations, or scientific computing. But in contrast to the semiautomatic air defenses that fueled software research, elite advisors expected that missile defenses would be controlled completely by computer. They analyzed defense with the disciplinary repertoire of physics, transforming an unpredictable future into physically determined battle scenarios. Human beings did not appear in these scenes. One of the chief goals of defense—protecting against an accidental launch or small surprise attack—continued to rest upon confidence in the computers that would control defense.

By the early 1960s, fallible computer programs were beginning to draw some attention. In July 1962, the *New York Times* featured a front-page headline: "For Want of Hyphen Venus Rocket is Lost."[144] The steering equations for the Mariner I, a spacecraft destined for Venus, had been coded incorrectly into the on-board computer. As the Mariner veered off course, NASA was forced to blow up the $18.5 million spacecraft. But such headlines were rare in the early 1960s. And in any case, elite scientists understood software not as computer code, but as the process of managing computers.

As we will see in the next chapter, the triumph of physical research and development carried an unforeseen irony. At the same time that scientist-advisors were helping improve radars and missile interceptors, they were finding reasons to argue that even the best missile defense technology should never be deployed. Yet, as the Army incorporated better physical components into its proposals for missile defense, deployment became virtually inevitable. By the time physicists began to note the limitations of software, the missile defense program was moving forward with a momentum all its own.

4   "No Technological Solution"?

The advantage of offense over defense . . . is the advantage of people working many years to try to develop penetration aids over a computer which must solve the problem in a matter of a few minutes. People are really smarter than computers. Computers do things faster. But planners who work on penetration aids can succeed, and can succeed with relative ease. . . .

—Herbert York, 1963 congressional testimony

In the mid-1960s, Herbert York, Jerome Wiesner, and other well-placed physicists began to advance a strange argument: the world would be safer without missile defenses. It was a curious reversal. As we have seen, these physicists advocated limited defenses in the late 1950s, viewing them as a form of arms control. Why did they come to see defenses as a threat? What kinds of reasoning guided recommendations to limit defenses? And how did they make these arguments politically persuasive?

The curious argument of York, Wiesner, and their compatriots has received considerable scholarly attention, in no small part because—as we will see in later chapters—it would become so influential. Historians and political scientists have shown that arguments against defense spread transnationally through forums like the Pugwash conferences, a series of meetings started by intellectuals seeking to reduce the threats of nuclear weapons.[1] In the "epistemic communities" model of influence, expert ideas are first formulated, and then selected by policymakers. However, this model does not explain why some scientists began to advocate limits to defense, while others advocated deployment, or why some arguments gained salience, while others fell by the wayside. Existing accounts tend to focus on dominant evaluations of missile defense, treating these as either a clear reflection of technical realities, or as a "pure" social construction.[2]

By contrast, this chapter shows how disciplinary training, political processes, and professional commitments all shaped scientists' arguments about missile defense. In McNamara's Defense Department, decisions about weapons procurement came to be justified primarily on the basis of physical and economic analyses. Systems analysts developed sophisticated ways of quantifying what physicists had argued from the mid-1950s onwards: defenses could be overwhelmed with countermeasures. Systems analysts showed that defenses would likely be more expensive to develop and deploy than to defeat, and persuaded McNamara that defense was a losing game.

However, physical and economic arguments left plenty of room for disagreement about whether or when a defense should be deployed. How rapidly could a defensive system be deployed? Sure, the Soviets *could* incorporate penetration aids into their arsenals—but *would* they? Was the U.S. offensive arsenal a sufficient deterrent, or might the Soviets actually contemplate a first strike? By the mid-1960s, scientists and engineers were divided on these questions.

Scientists and engineers most closely associated with the Defense Department grew enthusiastic about improved technological capabilities, and optimistic that defenses could be deployed quickly and inexpensively, before the Soviets incorporated penetration aids. They feared a first strike, and believed that the United States should prepare to defend cities. Sure, it would be expensive. But they believed the nation needed to stay ahead of the arms race.

By contrast, most of the president's science advisors expected that defenses would always lag offenses. As both the United States and the Soviet Union developed offensive weapons more rapidly and cheaply, they concluded that missile defenses were unlikely to facilitate an agreement to limit offenses. On the contrary, they expected that the Soviet Union would certainly develop penetration aids and deploy more weapons to overcome a defense. Hence, missile defense would only fuel an arms race, while leaving both sides vulnerable to nuclear annihilation. Most of these physicists— individuals who were more closely associated with civil society than the Defense Department—felt that the arms race was taking valuable resources away from domestic problems, and that it needed to be curbed. Still worse, they emphasized a defense of cities would be worthless without fallout shelters and other developments that would distort civil society.

Though most of the president's science advisors opposed city defenses, they believed that a defense of missile silos might eventually contribute to deterrence. The most notable exception to this rule was Herbert York. As early as 1960, Herbert York concluded that missile defenses faced not only *physical* and *economic* limits, but also fundamental *informational* limits. He reasoned that with only ten or fifteen minutes of warning, a defender could not gain enough information to reliably stop a complex, carefully planned attack. Though programmers would toil for years in advance, York felt that their chances of anticipating and countering the attack correctly were very low. He argued that there would be "no technological solution" to the arms race.

This was not an easy sell. How could policymakers refuse to defend their constituencies? As we will see, physicists made their arguments persuasive by framing them in terms of the dominant political concerns of the day—military spending and domestic priorities. At the same time, public debate opened up space for a long-marginalized argument, one based on common-sense reasoning about complexity and the limits of information technology.

1 Why Not Defend Ourselves? Technology, Politics, and the Arms Race

1.1 No Technological Solution?

The earliest arguments against deploying defenses emerged during debate over a treaty to limit nuclear testing. As we have seen, Eisenhower took the advice of Wiesner and other presidential science advisors when he entered a 1958 moratorium on testing. However, scientists and engineers associated with the nuclear weapons laboratories opposed the moratorium, often highlighting the ways that testing could contribute to missile defense research and development. When the United States and Soviet Union both resumed nuclear testing in 1961, they conducted several tests related to missile defense. Tests of missile interceptors also became an important part of the saber rattling that reached an all-time high in 1962.[3]

The Cuban missile crisis in October 1962 brought Kennedy and Soviet premier Nikita Khrushchev slowly back to the negotiating table, and in July 1963 they began to discuss limits on nuclear testing. In just twelve days, the Soviet and U.S. delegations drafted a treaty that would ban nuclear testing in the atmosphere, space, or underwater. On August 5, 1963, secretaries of state from the United States, Britain, and Soviet Union signed the treaty.[4]

That same month, the Senate Foreign Relations Committee opened hearings to discuss whether the Congress should ratify the treaty. Scientists at the nuclear weapons laboratories testified that nuclear testing was critical to resolving uncertainties associated with defenses. For example, John Foster, the director of Lawrence Livermore Laboratories, noted that missile defenses were "extremely complex" and argued that "simpler systems" had failed "when proof-tested in environments which are far better understood than that of a hostile nuclear situation."[5] Similarly, Edward Teller called missile defense "the most complex military operation" he'd ever known, and emphasized that it would be "most hazardous" to develop defenses without "proper and complete experimentation," including atmospheric testing.[6]

By contrast, most of the scientists and engineers associated with the President's Science Advisory Committee argued that the most important uncertainties associated with defense would not be resolved by further testing. Harold Brown, a former committee member who was serving as director of defense research and engineering, emphasized that the "most important" challenges for missile defense lay in the "design of the radar, missile, and data-processing systems."[7] PSAC member George Kistiakowsky and other physical scientists made similar points.[8]

Although scientists disagreed over the importance of nuclear testing, almost all agreed that the development of missile defenses was important to national security. The most notable exception to the rule was Herbert York. He argued that concerns about missile defense were "misplaced," and that the United States should instead focus on developing offenses that could penetrate Soviet defenses, because "the offense will always, and by a large margin, have the advantage over the defense."[9] York explained that the advantage of offense was one of timing and information, "the advantage of people working many years to try to develop penetration aids over a computer which must solve the problem in a matter of a few minutes."[10]

Senators readily accepted York's arguments about the limits of computers. When York noted that "people are really smarter than computers," William Fulbright, chairman of the Senate Foreign Relations Committee, replied: "They can be more original, I take [it], too, than computers."[11] Nonetheless, the committee focused less on computers than on a deeper argument running through York's testimony. York argued that the Soviet Union and United States were building such large nuclear arsenals that defenses could not possibly keep up, leaving each increasingly vulnerable: "This dilemma of steadily increasing military power and steadily decreasing national security has no technical solution. If we continue to look for solutions in the area of science and technology, the result will be a steady and inexorable worsening of the situation."[12]

York's remarks inspired arms controllers far and wide. Even before the Senate ratified the limited test ban treaty in September, the editor of *Science* magazine, Philip Abelson, wrote to York with "admiration" for his statement and invited him to prepare an article of his choosing.[13] But York requested a "possible rain check" on any publication. He hoped that the partial test ban would "be followed by a second and more significant step," and suggested that he might publish an article when "things begin to warm up in this area."[14]

What really heated things up was the 1964 presidential race between Republican presidential nominee Barry Goldwater and the Democratic incumbent, Lyndon B. Johnson. Goldwater quickly earned himself the reputation of being loose with nukes. Though Goldwater's campaign slogan was "In your heart you know he's right," Democrats countered: "In your heart you know he might."[15] Gerard Piel, the publisher of *Scientific American* and a long-time nuclear arms controller, sought out the article that York had alluded to the previous year.[16]

York had just left his position as chancellor of the University of California, San Diego, and was vacationing with his family in Mexico, when his secretary called to tell him that Piel wanted the article. York recalled feeling "especially lazy," so at first he refused. But his secretary persisted, explaining that Piel considered it "urgent" to publish the piece before the election. So York passed the buck, telling her that "if Jerry Wiesner will sign off" on the draft article, "they can have it." Wiesner did "sign off," and in October of 1964 *Scientific American* published a hastily edited and illustrated coauthored article.[17]

The first few pages of the article argued that the United States did not need to continue nuclear testing to develop more effective offenses, because it would gain more strategic advantage through improved missile accuracy than through larger bombs. Similarly, nuclear testing would not much improve defenses, because the real challenges lay in issues such as radar, guidance, and computing. The article continued by noting many reasons "for believing that defense against thermonuclear attack is impossible."[18]

On the eve of attack the offense can take time to get ready and to "point up" its forces; the defense, meanwhile, must stay on the alert over periods of years, perpetually ready and able to fire within the very few minutes available after the first early warning. The attacker can pick its targets and can choose to concentrate its forces on some and ignore others; the defense must be prepared to defend all possible important targets. . . .[19]

They noted that these uncertainties could not simply be resolved by faster computers.

Furthermore, because the "designer of the defensive system . . . cannot begin until he has learned something about the properties and capabilities of the offensive system, "the defense must start the race a lap behind."[20]

Finally, they noted that defenses would only be effective with fallout shelters: "the logical next step is the live-in and work-in blast shelter leading to still further disruption and distortion of civilization," and "a diverging series of ever more grotesque measures."[21] They concluded that the problems of national security had "no technical solution," warning: "The clearly predictable course of the arms race is a steady open spiral downward into oblivion."[22]

Piel used the article strategically, reprinting hundreds of thousands of copies and flooding the mailboxes of major newspapers and magazines, government offices, diplomats, and policymakers.[23] He succeeded in raising quite a stir. The *Washington Post* ran a front-page story noting the implications of the article for the Goldwater campaign, and several other major papers ran independent stories. The *New York Times* even endorsed the argument in an editorial headlined "Spiral to Oblivion."[24] One week after Johnson defeated Goldwater, Piel sent clippings to Wiesner: "I repeat that we were delighted to be the vehicle for delivery of this warhead."[25]

1.2   A "Crazy" Proposal

But could there be "no technical solution" to the problems of national security? Many scientists and policymakers immediately objected. As a long-time advocate of defenses, Paul Nitze drafted a letter for *Scientific American,* arguing that the article overemphasized the dangers of the arms buildup, while neglecting the need for the United States to stay in the race with the Soviet Union. But he sent the article to York first, and ultimately decided to stay out of the public debate: "Too many basic and sensitive issues are involved to make it profitable for someone in the government usefully to do so." Nonetheless, he chastened York: "I hope you and Gerry [Piel] . . . won't leave others, as you did me, with the impression that science is to be used as the tool of personal advocacy rather than as a basis for judgment."[26]

As this suggests, York's and Wiesner's article raised questions about the role of scientists in policymaking. Some physicists felt that they had a special role to play because technology *did* offer solutions to national security. As one sympathetic physicist pointed out to York, seismic detection technology helped verify the nuclear test ban.[27] And indeed, the same engineer who managed ARPA's work on seismic detection, Jack Ruina, was also the first to argue that the United States and Soviets should agree to limit defenses.

The idea first emerged as a joke. In 1962, around the time that Khrushchev bragged that the Soviets could "hit a fly" in space, the President's Science Advisory Committee met to discuss missile defense. Ruina recalled chatting with Wiesner and Hans Bethe at a coffee break:

Jerry, as a joke, said . . . "Why are we building this stupid system? We're only building it because the Russians are building this system. Why don't we tell the Russians, 'You stop building yours and we'll stop building ours.'" . . . And people laughed. And I said, well why is that such a joke, why don't we do that?[28]

Ruina surely recognized that limits on defenses could be verified without invasive on-site inspections, because any defensive system would depend upon very large, exposed phased-array radars that could be viewed from space. He soon formulated a serious proposal for ABM limitation and took it to his boss, Harold Brown:

Harold Brown said, ". . . talk to [Deputy Secretary of Defense] Ross Gilpatric. He'll think it's a good idea. But I think it's crazy. Not technically, but politically it's crazy."

So sure enough, I made a point of seeing Ross Gilpatric. . . . And Gilpatric, in his gentlemanly way, looked and he said, "That's a very interesting idea. You ought to pursue that further."[29]

Ruina did pursue it further: at a 1964 Pugwash conference, he delivered an oral presentation on the risks of missile defense. He recalled the Russian response: "This is such a crazy idea the interpreters must have done a bad job. So why don't you put it in writing?"[30]

Ruina wrote a paper that evening, and Murray Gell-Mann, a Nobel Laureate in physics, edited it the next morning. They explained that because "ballistic missiles are cheap and can do devastating damage, while defence is costly with very limited effectiveness," any defense would likely be overcome by adding additional offenses on the other side. Thus, "populations would still in essence be kept as hostages," yet "the cost to each side would be great and the added dangers significant. Uncertainty and suspicion would be increased."[31] They thus reached a "surprising" conclusion: missile defense deployments "may not be desirable."[32]

Ruina recalled that "the Russians thought it was the stupidest idea they'd ever heard of, not defending yourself."[33] Nonetheless, they agreed to think it over. Meanwhile, Wiesner and Gilpatric took the idea up to more formal channels. In a November 1965 report written for a White House Conference on International Cooperation, they called for a three-year moratorium on deployment of defenses by both the United States and Soviet Union.[34]

The Johnson administration did not immediately embrace recommendations to limit defenses.[35] Nonetheless, McNamara began to consider the advantages of limiting defenses after his systems analysts formalized Ruina's basic point: defenses would cost more to build than to defeat.

In May of 1963, the director of defense research and engineering, Harold Brown, commissioned a cost-benefit analysis of a "damage-limiting" strategy—wherein the United States would respond to a Soviet first strike with missile defenses, civil defenses, and retaliatory weapons, in an effort to limit the damage at home. This analysis showed that limiting damage was a losing game. To save about half of industry and 60 percent of the population, the United States would be forced to spend $3.20 on defenses for every $1 spent by the Soviet Union on offenses. If the United States wished to protect 70 to 80 percent of its assets, it would have to spend between \$10 and \$20 on defenses for every dollar the Soviets spent on offenses. On the bright side, the analysis suggested that the United States need not worry about the Soviet defenses, since it would be relatively cheap for U.S. offenses to overwhelm defenses and inflict considerable damage on the Soviet Union.[36]

1.3   The Technopolitics of Terminal Defenses

McNamara was impressed by these conclusions. Nonetheless, with billions of dollars invested in defensive research and development, he faced pressures to find some use for defenses. In 1964, he asked ARPA and the Office of Systems Analysis to study defense from a small attack. With phased array radars and new information about penetration aids—both developed under ARPA's sponsorship—the study concluded that high altitude defenses might be able to protect a relatively large area from just a few warheads and decoys.[37]

Pressures for deployment grew on October 16, 1964, when China detonated its first atomic bomb. Strategic analysts began to worry that China might eventually be able to "blackmail" the United States by holding a city hostage. Furthermore, the continued development of a missile defense system around Moscow, combined with the U.S. failure to incorporate penetration aids into its offensive weaponry, left some military planners anxious: what if the Soviets' primitive defense had given it strategic superiority?[38]

In June of 1965, McNamara received the Army's proposal for "Nike-X against Light Attack," a system that would include two main components. First, long-range interceptors armed with high-yield nuclear weapons would destroy warheads and decoys at high altitude, defending large areas. Second, fast "Sprint" interceptors armed with lower-yield nuclear weapons would intercept any warheads that leaked through the area defense, providing a terminal defense of six cities: Los Angeles, New York, New Orleans, San Francisco, Seattle, and Chicago.[39] McNamara sent the Army's proposal to the President's Science Advisory Committee.

In August 1965, Marvin ("Murph") Goldberger, a physics professor at Princeton and the chair of PSAC's Strategic Military Panel, led a review of the Army's proposal. Goldberger was among several newcomers to the panel, including Sidney Drell and Richard Garwin. While the founding members of PSAC had taken leading roles in wartime research, Goldberger, Drell, and Garwin earned their doctorates in 1949, and in the wake of Sputnik helped start a secret group of Defense Department advisors, the "Jasons."[40] Goldberger later recalled that "Jason provided a kind of training ground for young people," and that PSAC initially "used Jason as a farm camp."[41]

The younger cohort of physicists continued PSAC's tradition of broad-ranging studies. When Goldberger hand-drafted the committee's report, he noted: "It is virtually impossible to separate the technical and political arguments for or against missile defense and we shall make no attempt to do so here. The political ones are undoubtedly overriding and we won't withhold our advice on these."[42]

Goldberger acknowledged that defense might seem appealing given technical advances and the "emergence of China as a nuclear power . . ."[43] But his panel expected that the Army's initial proposal, to provide six cities with terminal defenses, would surely expand because the blackmailer could always target city number seven:

While in some vague way New York, Chicago, and Los Angeles are of greater "value" than St. Louis, Cincinnati and Cleveland, the residents of the latter would hardly agree and the country would be reluctant to have them disappear.[44]

Thus, the panel would only consider a high altitude area ("umbrella") defense designed to defend against a small attack (i.e., from China). While it noted several potential advantages of area defense, it also noted risks. The U.S. allies might "like us to buy them an umbrella," but this would be "prohibitively expensive." The system could not stop attacks from Submarine Launched Ballistic Missiles or aircraft, and thus might encourage a "Maginot-line" mentality. Since China had not yet developed ICBMs, a defense against that nation might ironically enhance its prestige. And since even a small "anti-Chinese" defense would be able to confuse a Soviet attack, it would encourage a Soviet offensive build up and "a further arms spiral."[45] Goldberger mused:

. . . the panel seemed to find the area defense concept technically rather attractive but in various ways expressed reservations about actually buying the system. There is a kind of finality about taking a step which could make war somehow acceptable or bearable or survivable. . . . The current arms configuration in the world is grotesque and unstable. . . . To continue monotonically on a course of actions toward the Chinese which is exactly the one followed in response to the Soviet threat seems wrong.[46]

Indeed, the panel was so opposed to a new arms race that it briefly considered the "drastic" possibility of "removing the Chinese threat now."[47] Goldberger noted: "I am personally very depressed about any deployment, ever."[48]

The panel's final report emphasized that the "far-reaching military, economic, and political consequences" of deploying defenses "may be to the long-term net disadvantage of the United States."[49] Most significantly, it noted that the Army's proposal for a "thin" defense would inevitably expand into a "thick" defense against a much heavier attack, escalating an expensive arms race.

Thus, the panel recommended that the Army's proposed system not be deployed. Instead, it asked that the Army develop "a simplified area defense system which would be relatively inexpensive, use off the shelf components, and be rapidly deployable."[50] Eliminating the "terminal" elements of a city defense would not only limit costs and time-to-deploy—it would limit the system's potential to grow and to fuel the arms race.

1.4   Technological Imperatives and the Arms Race

McNamara accepted these recommendations in the fall of 1965. He deferred deployment in his proposed budget for 1967, and directed the Defense Department to cut funds for technologies that aimed to defend "against a major Soviet threat."[51]

While McNamara embraced PSAC's recommendations for restraint, other scientists favored a near-term deployment. Perhaps most notably, John Foster came to the DDR&E from the directorship of Livermore Laboratories—as had York and Brown—but his professional identity was far more rooted in military research. York and Brown worked actively with the President's Science Advisory Committee, a group that was dominated by university scientists, reported to the president, and tended to evaluate technology in terms of its broad policy implications. By contrast, Foster retained closer ties with the Defense Science Board, a group dominated by scientists in industry and government that reported to the DDR&E and tended to evaluate technological capabilities in more narrow terms.[52]

In early 1966, Foster asked the Defense Science Board to study possibilities for defense against the Soviet Union. A task force was soon convened under the chairmanship of RAND physicist Richard Latter. Seeming to follow PSAC's recommendation for a rapidly deployable area defense, the Nike-X Project office provided Latter's group with four deployment options that eliminated the "Sprint" interceptors required for terminal defenses. However, two of the options retained "missile site radars" designed to defend local targets, suggesting that the Army wanted to retain the option of city defense.[53]

While most members of PSAC wanted to avoid terminal defenses of cities—and thereby to avoid the system's growth and the acceleration of the arms race—the Task Force explicitly sought a system that could keep up with the race. The panel explained that a "vast R&D program" had produced "a substantial body of defense technology" which should be rapidly deployed.[54] In particular, the group recommended deploying terminal defenses at an estimated cost of between $6 and $8 billion for construction and five years of operation. They acknowledged that the system's long-term effectiveness could only be maintained at great cost, but still favored proceeding with the system.[55]

The panel's recommendation to deploy was partly rooted in fears of a first strike. Whereas PSAC expected that the U.S. offensive arsenals would deter a Soviet or Chinese attack, Latter's group imagined that the deterrent might fail. They noted that since the Soviets were in the process of deploying defenses, but the United States had not yet deployed penetration aids, "we will be unsure of our damage-inflicting capability."[56] If the United States could no longer deter an attack, the panel warned, it might "no longer be able to deal resolutely with Soviet provocation."[57]

The panel's solution was to limit damage. They concluded that if the Soviet Union attacked with 1,000 missiles, their recommended defense would reduce fatalities from 100 to 145 million Americans, to as few as 50 million Americans. If the Soviet Union launched 250 unsophisticated missiles, and the defense system operated perfectly, the United States might reduce fatalities from 55 to 60 million Americans, to 10 million Americans.[58]

Did the analysts really believe that the United States could or should deal more "resolutely" with the Soviet Union, knowing that it stood to lose 10 million Americans? McNamara had already decided that this was not an acceptable level of damage, and resisted committing to defenses.[59] Nonetheless, scientists such as Foster and Latter applied increasing pressure to deploy.

In fact, the entire DSB report was remarkably optimistic. The Nike-X Project Office estimated that some costs would increase by up to 40 percent to facilitate rapid deployment, but the report dissented without explanation: "The Task Force does not believe the 40% but considers that the increase will be substantially less."[60] Estimates of system effectiveness assumed that the system could be deployed before the Soviets responded, and that it worked ideally. The report also acknowledged that the "most difficult aspect of the system to design and check out will probably be the software," but argued that this "can probably be solved through adequate simulation testing."[61]

2   Physicists as Advisors and Activists: The Making of a Public Debate

2.1   Deploying an "Anti-Republican" Defense

Defense Department scientists did not confine their optimism to classified studies. For example, in open Senate testimony, Foster expressed his view that a thin defense stood "a very good chance" of stopping a small attack.[62] When pressed, Foster acknowledged that millions of lives would be lost in a determined attack, but emphasized positive aspects of defense: "Millions of lives would be saved."[63] Although he did not outright oppose McNamara's policies, such statements contributed to broader pressures to deploy defenses.

By the mid-1960s, the Joint Chiefs of Staff presented McNamara with unified support for several weapons programs, including missile defense. Additionally, with the escalation of operations in Vietnam, the U.S. Congress was eager to reassert control of military policymaking. On April 29, 1966, the Senate Armed Services Committee authorized the Defense Department's entire appropriation request, including the Army's request for preproduction funds for missile defense—which McNamara and Johnson had rejected.[64]

This did not force Johnson's hand, but it worried him. Both McNamara and his deputy continued to oppose any defensive deployment. They believed that it was unnecessary and would inevitably become a "thick" defense costing at least $30 billion, with no net gain in national security. But they also felt that "a substantial majority" of the Congress favored deployment, and that pressures were likely to grow.[65]

In an effort to unify his administration, Johnson called McNamara and the Joint Chiefs for a special meeting in December 1966. McNamara carved out a compromise: the administration would include a contingency fund for preproduction of a thin missile defense in the 1968 budget, to be spent only if efforts to negotiate limits with the Soviet Union failed. That afternoon, McNamara got the State Department working on negotiations.[66]

In January 1967, McNamara prepared for a battle with Congress by calling together the Joint Chiefs, all the past and present presidential science advisors and all past and present Directors of Defense Research and Engineering. York recalled that he asked a seemingly simple question: "Will it work and should it be deployed?" The scientists and engineers unanimously agreed that a "thick" defense against a Soviet attack should not be deployed, and the majority agreed that proposals for a "thin" defense were not worth deploying either.[67]

Despite scientific support, the administration found it increasingly difficult to refuse deployment. The Army continued to press for deployment, and on June 17, China detonated its first thermonuclear weapon, prompting several members of Congress to attack the administration for its refusal to deploy a defense.[68]

Meanwhile, the Soviets did not seem eager to discuss limits on defenses. Johnson glimpsed a tiny window of opportunity in mid-June when the Soviet premier Alexei Kosygin announced that he would visit the United Nations in New York. The administration hastily arranged a half-way meeting place in Glassboro, New Jersey. But two days of meetings were dominated by issues such as Vietnam and the Arab-Israeli war. A short, disorganized presentation of McNamara's case for limiting defense failed to persuade Kosygin.[69] In a press conference at the United Nations, Kosygin emphasized that "the anti-missile system is not a weapon of aggression, of attack, it is a weapon of protection—it's a defensive system."[70]

A few days after the meetings at Glassboro, Johnson privately told McNamara that he had decided to deploy a missile defense system, both because he wanted to pressure the Soviets to resume negotiations, and because he was concerned that Republicans would revive the notion of a "defense gap" in the 1968 presidential elections. McNamara was left to decide upon a specific deployment proposal, and to rationalize it publicly.[71] McNamara soon drafted an announcement of plans to deploy a "Chinese-oriented thin ABM system."[72] In early September, McNamara sent a message to the Soviets, reiterating that the deployment was not directed at the Soviet Union.[73]

On September 18, 1967, Robert McNamara finally delivered a carefully scripted speech to a room of editors in San Francisco. He repeatedly emphasized that the problem with deploying an anti-Soviet defense was *not* cost, but rather the "penetrability of the proposed shield." He emphasized that his scientific advisors had recommended not deploying such a defense. He continued by emphasizing that it would be "insane and suicidal" for China to launch a nuclear attack on the United States or its allies. Nonetheless, he explained that since "conservative" strategic planning must consider even "irrational behavior," the United States had "marginal grounds" for a light deployment of missile defense. McNamara warned against the "mad momentum intrinsic to the development of all new nuclear weaponry," and issued a call to replace the "race towards armaments" with a "race towards reasonableness."[74]

McNamara's speech achieved its ironic goal: journalists recognized the deployment decision as a political hedge in an election year, an "anti-Republican" defense.[75] Many criticized the decision based on McNamara's own argument: defenses were likely to accelerate the arms race.[76] McNamara was so frustrated with the contradictions of the administration—not only on missile defense, but also in Vietnam—that he quit his post soon thereafter.[77]

2.2 Physics and New Modes of Political Engagement

Most journalists accepted the decision to deploy what came to be known as "Sentinel" as final.[78] However, the most heated public debate had yet to come. Members of PSAC not only disagreed with the decision, they felt betrayed by the way it was announced. McNamara had invoked their authority to oppose heavy deployment, but did not mention that they also opposed thin deployment.[79]

Hans Bethe—a renowned theoretical physicist, Manhattan Project veteran, member of PSAC's Strategic Military Panel, and winner of the 1967 Nobel Prize in Physics—was one of the first to speak out publicly. When Gerard Piel asked Bethe to speak at the Annual Meeting of the American Association for the Advancement of Science (AAAS) in December 1967, Bethe readily accepted. Richard Garwin and Marvin Goldberger joined Bethe in missile defense panel at the AAAS meeting. Garwin and Bethe soon teamed up to turn their symposium comments into an article for Piel's *Scientific American.*[80]

Garwin and Bethe drew skillfully upon physics and publicly available information to construct a devastating critique of missile defense. Although their inside experience was a useful guide, they drew upon non-secret information as they quantified at least 47 aspects of the defensive-offensive standoff, with 7 formulae and 12 detailed examples to illustrate the limits of a defense. This physical analysis provided unprecedentedly detailed support for the dominant argument against deployment: defense from the Chinese was unnecessary, while defense against the Soviets would be both expensive and ineffective.[81]

Garwin emphasized that the United States and the Soviet Union could "annihilate each other as viable civilizations with a day and perhaps within an hour."[82] Bethe showed how each side could maintain this capability against any foreseeable defense. For example, the offensive missiles could be hidden by decoys, chaff, and nuclear precursor blasts, which would effectively blind the defensive radars (figures 4.1 and 4.2).

Thus, they argued that while a light defense would "add little, if anything, to the forces that should restrain China indefinitely" from attacking, it would "nourish the illusion that an effective defense against ballistic missiles is possible," leading "inevitably to demands that the light system, the estimated cost of which exceeds $5 billion, be expanded into a heavy system that could cost upward of $40 billion."[83]

In conclusion, Garwin and Bethe underscored the "strong social and psychological effects" of such an escalation:

The nation would think more of war, prepare more for war, hate the potential enemy and thereby make war more likely. The policy of both the U.S. and the USSR in the past decade has been to reduce tensions to provide more understanding, and to devise weapon systems that make war less likely. It seems to us that this should remain our policy.[84]

Garwin's and Bethe's arguments first appeared in the March 1968 issue of *Scientific American* and resounded over the next several months, in major

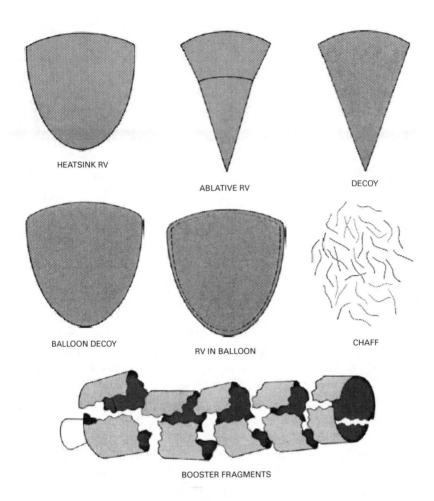

Figure 4.1
Offensive missiles could be hidden from radars using penetration aids that would reflect radar signals and confuse the attack. Reproduced with permission. Copyright © 1968 Scientific American, Inc. All rights reserved.

newspapers, the Pentagon, the scientific community, and the U.S. Congress.[85] Qualitatively, the arguments were not new, but the detailed analysis from individuals with such "impressive credentials," left the Pentagon anxious about defending its plans.[86] The Federation of American Scientists (FAS) noted that Garwin and Bethe went "farther in addressing some specific technical questions—such as radar blackout resulting from nuclear explosions—than other authors . . . in the open literature."[87]

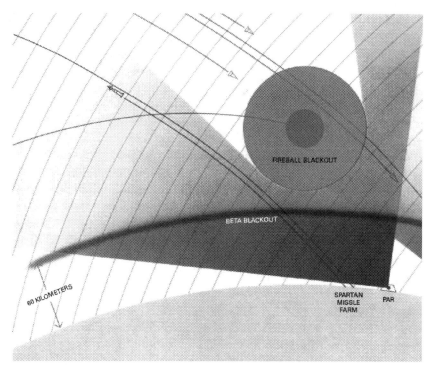

Figure 4.2
The heat of a nuclear fireball ionizes air over a relatively localized area, and the beta radiation it emits ionizes air over a much larger area. The resulting plasma attenuates radar, making it difficult or impossible to see the incoming attack. Reproduced with permission. Copyright © 1968 Scientific American, Inc. All rights reserved.

For many scientists, the missile defense deployment prompted questions about proper modes of political engagement. With protests against the Vietnam War rising, elite scientist-advisors found themselves in the cross-hairs of activists who objected to the Defense Department establishment. In fact, many of the elite advisors were also feeling disillusioned about the Vietnam War. For example, Goldberger and other members of the Jasons helped design technology that they hoped would reduce bombing in Vietnam, only to discover that the military used their technology as an "add-on" to increasingly destructive bombing.[88] Many of the Jasons quit. In March 1968, *Science* reported that George Kistiakowsky had "quietly severed his connections with the Department of Defense (DOD) because of his opposition to the administration's Vietnam policies."[89]

Although elites tended to abstain from public comment about "political" issues such as the Vietnam War, the Sentinel decision suddenly gave them a "technical" rationale for publicly opposing military policies. Jerome Wiesner explained: "I stopped working within the government and started to work outside. After the so-called thin system decision, I gave up trying to convince anybody in the government to make sense on the ABM, for I regarded that as basically a political decision."[90]

As elite physicists sought to generate and win a public debate about missile defense deployment, they found a brief alliance with more radical scientists. Perhaps most notably, in the fall of 1968, some MIT graduate students in physics began to organize a day of discussion about the relationship between science and the military. Initially conceived of as a scientists' "strike" against military research, the students drew in elite speakers by recasting it as a "teach-in." Hans Bethe was at the top of their list of invited speakers, in part because of his public opposition to defense, and they scheduled the day to suit him. Soon students at other universities began planning discussions for the same day. On March 4, 1969, scientists at 36 universities across the nation engaged in a highly publicized day of discussion about the relationship between scientists, politics, and social needs.[91]

Most elites argued that their association with the defense "establishment" was good because it provided them with technical knowledge. Hans Bethe began his address to a crowded room in Cambridge: "I believe that most of the audience here is against the ABM, and I believe that I am here to tell you why."[92] He summarized his arguments in *Scientific American*, and concluded by emphasizing the value of inside knowledge: "I could not have given you the arguments tonight . . . if I were not . . . an "in" man."[93] He continued:

. . . without the in men you probably would never have known that the antiballistic system is dangerous. After all, it saves lives, doesn't it? . . . That is the superficial view that anybody will have of such a weapon if there are no people who are inside, who study what such a weapon means and know the technological background of both the offense and the defense.[94]

Sidney Drell delivered similar remarks at Stanford, as did Herbert York at UCSD.[95]

2.3   The Making of a Public Debate

Scientists and engineers also sought to raise local, grassroots opposition to defenses. The Army insisted that stationing Sentinel battalions near cities would offer them special protection. However, scientists associated with several local chapters of the Federation of American Scientists used physical analyses to show that the Army did not need to locate its battalions near cities to provide the "thin" area defense. They accused the Army of attempting to deceive the public about its real goals, and called for open discussion before Congress.[96] They also raised local opposition by citing risks associated with stationing the interceptors near cities, such as an accidental nuclear detonation, increased fallout if the defenses were used, or the possibility that the Soviet Union might attack the defense first, turning cities into valuable targets.[97] By December of 1968, at least five chapters of the FAS reported local concerns about the ABM issue, in Cambridge, Chicago, Detroit, Los Angeles, and Seattle.[98]

Many city residents who opposed local battalions continued to view defenses as an important aspect of national security. As the *Chicago Tribune* reported: "Almost everyone agrees that a string of nuclear anti-missile bases spread across the country is important to our national defense. 'But why build one here?'"[99] Similarly, *The Atlantic* suggested that "the intensity of the anti-ABM sentiment was a fluke, provoked by good old American feelings about real estate."[100] By contrast, most of the scientists who opposed Sentinel focused on the arms race and its relation to a much broader debate—about the scientists and policymaking, technology and war, domestic needs and military spending.

Arguments against missile defense grew only slowly in influence, in part because defensive deployments remained marginal relative to the war in Vietnam, urban riots, poverty, and presidential elections. Missile defense only made the front page of the New *York Times* three times in 1968. Indeed, Johnson successfully prevented missile defense from becoming a major election issue. By October 1969, missile defense seemed a fait accompli (see figure 4.3). The *New York Times* reported "an unusual secret session" in which the Senate "overwhelmingly rejected . . . what was probably the final effort to delay deployment" of Sentinel.[101]

Unexpectedly, the closed session was a turning point, as senators mobilized scientists in new ways. Senator William Fulbright, the chair of the Foreign Relations Committee, criticized the Armed Services Committee for following the president's recommendations to deploy Sentinel without hearing testimony from Wiesner and other scientists who opposed deployment. Senators who favored deployment rejoined that scientists such as Wiesner were making "political" rather than "scientific" arguments, prompting considerable debate. Finally, Philip Hart of Michigan, a leader of a Senate coalition against Sentinel, asked whether it wouldn't be wise to have "an opinion available so that we could evaluate to what degree it is scientific and to what degree it is political?" Proponents of deployment conceded the point. Fulbright jumped on the opportunity by asking, and winning, "unanimous consent that these communications be included in the open Record as well as the secret record."[102]

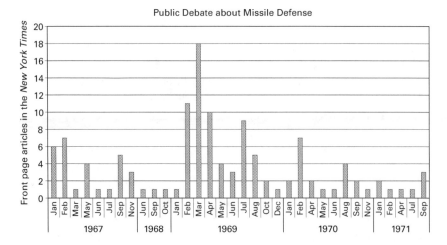

Figure 4.3
The prominence of antiballistic missile defense debate over time, as measured by front page headlines in the *New York Times*.

The public record appeared on November 1, and the December issue of *Science* magazine ran a prominent headline about the Senators' request for "outside scientific advice."[103] The chairman of the Federation of American Scientists immediately wrote to the chair of the Senate Armed Services Committee, offering to supply witnesses. Similarly, Wiesner and York requested the opportunity to testify.[104]

The political winds were also shifting, for Republican Richard Nixon narrowly won the election over Democrat Hubert Humphrey. Nixon was a well-known anti-Communist who favored city defenses, and could be expected to continue Johnson's program. No doubt it was easier for Democrats to oppose Nixon than to oppose Johnson. Both the Senate Armed Services Committee and the Senate Foreign Relations Committee began to plan hearings with non-administration scientists.

2.4 Catastrophic Failure? The Emergence of a New Argument

When the hearings opened on March 6, 1969, Sentinel opponents were initially disappointed. Gerard Smith, director of the Arms Control and Disarmament Agency, argued that beginning a "thin" deployment would have "little, if any, impact" upon the Soviets' interest in negotiating arms limits.[105] Bethe, a Nobel laureate, repeated the arguments of his increasingly cited the *Scientific American* article, but also expressed support for deploying defenses around missile silos "at the appropriate time," because this would "stabilize the strategic situation rather than the opposite."[106] Like the other two physicists who testified that day—former director of defense research & engineering Daniel Fink and former ARPA director Jack Ruina—Bethe agreed that missile defenses might be useful in some circumstances.

Significantly, arguments in favor of such "hard point" defenses rested on physical and economic analyses that were highly idealized. Bethe began his testimony by expressing optimism: "I have no criticism of the components of the Sentinel system. . . . I believe they are well designed and I assume they will work as designed."[107] Bethe and others recognized that even a perfectly functioning system could not protect populations, but believed that such a system might provide strategic stability by protecting missiles.

A second session of testimony brought out a far deeper objection to defense. On March 11, 1969, Herbert York took the stand and introduced "a technical problem that pertains to all the forms of ABM so far proposed, but which unfortunately is not so simple to discuss nor so easy to quantify as those brought to your attention last week."

Any active defense system such as the ABM must sit in readiness for 2 or 4 or 8 years and then fire at the precisely correct second following a warning time of only a few minutes. This warning time is so short that system's designers usually attempt to eliminate human decision makers. . . . Further, the precision needed for the firing time is so fine that machines must be used to choose the precise instant of firing no matter how the decision to fire is made . . . the trigger of any ABM, unlike the trigger of the ICBM's and Polarises, must be continuously sensitive and ready, in short a 'hair trigger' for indefinitely long periods of time.[108]

This led to "contradictory requirements" for the defense: "a hair trigger so that it can cope with a surprise attack, and a 'stiff trigger' so that it will never go off accidentally."[109]

Thus York expressed "the gravest doubts" about any missile defense system, and predicted a failure: "I am not here talking about some percentage failure inherent in the mathematical distribution of its distances, nor statistically predictable failures in system components but, rather, about catastrophic failure in which at the moment of truth either nothing happens at all, or all interceptions fail."[110]

Although York had not mentioned computers, senators made the connection. Albert Gore, Sr. explained that York had given him "an uneasy feeling" about "the dependence of our national security . . . upon a hair-trigger computer rather than a decision by some chosen official."[111] He continued: "Those of us who have had even limited experience in programming business needs into a computer system have only to find that some small but entirely unexpected variation from the pattern throws the whole computer report out of kilter."[112] Indeed, as we will see in the next chapter, programming glitches were becoming increasingly visible.

Gore next asked George Kistiakowsky and James Killian, York's friends and fellow witnesses, to elaborate. Kistiakowsky emphasized that an accidental launch could be consequential: everyone who looked up would be permanently blinded, and there would be hazards from fallout. But because the missile would be launched on only a few seconds or minutes of warning, it would be impossible to get the president involved. Instead, the system would rely on a very fast computer, which is "really a very stupid thing . . . it is the programmer who must be clever."[113] So the decision to fire would be made "years ahead by computer programmers or by a junior officer on duty at the instant of the event happening."[114] Still, Kistiakowsky felt that "there is no doubt that the human commander would retain the opportunity for overriding the computers, and not to have that is unthinkable."[115]

York went further: "I agree that it ought to be unthinkable, but unfortunately it is not."[116] He explained:

That is exactly the direction we are headed in. We are not talking about the design of the ultimate weapon, but the ultimate absurdity. The machines are getting so as to complexities and time scales to where there is not any room for a human being if we keep on going the way we are going [sic].[117]

York's and Kistiakowsky's remarks captured the imagination of journalists, policymakers, and other observers. Although they were not the first Sentinel critics to highlight the challenges associated with complex weapons systems, they were the first to tie these concerns explicitly to the arms race.[118] The *New York Times* noted that the scientists had "introduced a new argument" into the debate by noting that "the power to make certain life-and-death decisions is inexorably passing from statesmen and politicians to more narrowly focused technicians, and from human beings to machines."[119]

The publicity surrounding York's testimony prompted new conversations among scientific and military elite. The *Baltimore Sun* juxtaposed York's comments with those of General Alfred Starbird, the director of the Sentinel program. Starbird had stated that "launch can occur only when a threat of actual attack appears and then only on the basis of proper authority."[120] York sent the general a copy of the article and asked for a conversation.[121]

Starbird replied: "I feel your use of the term 'hair-trigger' is both unfortunate and ill-advised. It is not a fair representation at all . . . considering the various safeguards that will be incorporated."[122] He noted that York's comments had "led to many queries as to whether computers will make the decision to engage a suspicious object," adding: "I think we can look forward to repeated claim [sic] that a former Director of Defense Research and Engineering, who should know, has said that a computer will make the decision. Whether they will believe me when I answer that it will not, I am not certain."[123]

3   Conclusion: Public Debate and the Emergence of New Arguments

Thus, the public debate opened up the space for new kinds of arguments about missile defense. In the Defense Department of the 1960s, the dominant argument against defense came to rely upon highly esoteric physical and economic analyses that idealized technology. These analyses showed that a defense of cities could always be penetrated and would fuel an arms race, but suggested that defenses of missile silos might eventually be cost-effective and would stabilize the arms race. By contrast, public debate brought out a deeper concern: the very fast response needed for defense would lead to an unprecedented reliance on complex, failure-prone computers. In York's view, the need for any defense to have a "hair-trigger" response would encourage an increasingly dangerous, reactive, nuclear stand-off.

Growing Congressional opposition to defense was just one reason that Nixon began to reconsider Sentinel as he entered office. In January 1969, the outgoing deputy secretary of defense for systems analysis, Alain Enthoven, recommended that Nixon delay or cancel Sentinel. The Soviet and Chinese threats had not developed as rapidly as predicted and domestic political pressures for a new strategic weapon had diminished, while the Army's costs appeared to be rising with little hope of accountability. Thus, Enthoven argued, the system would not "provide full value for what it costs."[124]

Additionally the Soviet Union seemed increasingly interested in negotiating limits to defense. As Matthew Evangelista has discussed, Soviet scientists began to embrace the arguments of their American counterparts around the same time that McNamara decided to deploy.[125] Near the end of June 1968, the *New York Times* published an essay by the prominent Soviet dissident and nuclear physicist Andrei Sakharov, which pointed to Bethe's and Garwin's arguments against defense, and advocated international cooperation.[126] When Moscow and Washington signed the nuclear nonproliferation treaty on July 1, 1968, they issued a joint statement announcing plans to enter talks on limiting both offense and defense "in the nearest future."[127] Though the Soviet invasion of Czechoslovakia in August 1968 delayed talks, Nixon had good reason to eventually resume negotiations. Mired in the mess of Vietnam, he needed a foreign policy success.[128]

At the end of January 1969, the new secretary of defense, Melvin Laird, embraced Sentinel in the interest of negotiating from a position of strength and maintaining strategic "sufficiency."[129] But just one week later, he announced that the Sentinel program would be paused for a month-long review.[130] By late February, rumors began to circulate that Nixon might reorient defenses from protecting cities to protecting missiles and bombers, thus eliminating several major sources of public opposition.[131] On March 14, three days after York's appearance in the Senate, Nixon finally announced that he would be deploying fewer missile defense battalions—only twelve—and would deploy them near Minuteman silos. He justified the new "Safeguard" system as something that would enhance deterrence while also defending cities from a light attack.[132]

Nixon's announcement was carefully calculated to appease public opposition, while maintaining flexibility about what, exactly, the United States would deploy. By moving many of the battalions away from cities, Nixon hoped to provide a "thin" defense of some cities without raising too much local opposition to nuclear deployments. He also hoped to eliminate suspicions that he was preparing for a "thick" defense, and emphasized that the new deployment should not spark a Soviet reaction. At the same time, by defending Minuteman silos, he hoped to enhance deterrence—a goal that even physicists such as Hans Bethe might support.[133]

But physicists immediately objected that the defensive technology was not suited to this new mission. Furthermore, as we will see in coming chapters, Nixon's shift represented a broader shift in nuclear strategy—one that would leave the United States increasingly reliant on "hair-trigger" computers." Proponents of defense would continue to maintain confidence in the reliability of the system, while many opponents returned to York's comments about the risks of complex, hair-trigger systems.

Who, then, would settle the debate about the reliability of complex computers? None of the dominant defense experts in the 1960s were computer experts. Yet in the spring of 1969, several computer experts were beginning to articulate the challenges of complex software and to express their concerns about the risks of software for the Safeguard system. To understand how they framed their arguments, and how these were taken up into the Safeguard debate, we must turn to the birth of "software engineering" in 1968.

5   What Crisis? Software in the "Safeguard" Debate

Since facts about cost, timeliness, and initial performance of large, complex [computer] systems tend to reflect adversely upon those in positions to write authoritatively . . . the facts are not fully reported in the open literature. Most of the facts I have come across are well remembered but not well documented experiences.

—J. C. R. Licklider to Jerome Wiesner, March 24, 1969

In early March 1969, Jerome Wiesner went digging for dirt on the risks associated with large and complex systems—excessive costs, slow development, errors, and catastrophes. Wiesner was organizing a study for Senator Edward Kennedy (D-MA), a prominent opponent of missile defense deployment, and he wanted a chapter on reliability. Wiesner turned to his long-time colleague, J. C. R. Licklider, and struck gold. Within a day, Licklider did some "brain-picking" of colleagues and returned a list of 10 complex computer systems that had proven problematic.[1] But after 17 more days of digging, Licklider reported only "anecdotal evidence."[2] He concluded that even "with power of subpoena . . . it would be difficult or impossible to develop truly authoritative data" on the risks of missile defense software.[3]

Why was it so difficult to develop an authoritative critique? As we have seen, computer glitches were becoming common knowledge by the late 1960s—so much so that when Herbert York warned of a "catastrophic" missile defense failure, senators immediately thought of fallible computers. Talk of computer glitches carried a commonsense authority. However, computer experts sought a different kind of authority, akin to the laws of physics. After all, Bethe and Garwin had used physical laws and codified data to publish the most widely cited critique of missile defense. It was in the arena of formal analysis that computer experts came up short. What laws governed complex software? What qualified anyone to analyze it? And how could computer experts influence the missile defense debate?

This chapter shows how computer experts wrestled with these questions in the late 1960s, and to what effect. Significantly, the missile defense debate heated up just a few months after a small, elite group of computer experts sought to establish the "theoretical foundations and practical disciplines" of a new field: "software engineering." As historians of computing have shown, several participants in the world's first "software engineering" conference advocated the new field as a means of resolving a "software crisis."[4] This account extends these analyses by examining the origins of the putative crisis, dispute over its meaning, and the implications of that dispute for professional engagement in a highly public debate.

As we will see, many participants in the world's first software engineering conference blamed the crisis on a conflation of software research, development, and production. All too often, software developers discovered the need for more research only after the products behaved unreliably in real-world use. Most of the computer experts who opposed the missile defense system were barely aware of "software engineering," but they were concerned about a similar conflation. They argued that without operational testing, missile defense software would forever be in a stage of research and development, and never a trustworthy product. They pointed to myriad examples of software projects that only became reliable after real-world use revealed the need for further research and development.

However, many computer experts objected that progress in the art and science of programming would eventually triumph over reliability problems. Without formal proof or inside evidence of problems, they insisted that reliable missile defense software could be developed. In the late 1960s, computer experts had no formal, codified, or quantitative way of analyzing the risks of software development—no disciplinary repertoire. Without any consensus on the challenges of complex software, policymakers brushed the issue aside. They presumed that with sufficient engineering effort, software would become reliable.

1   Software Research, Development, and Production

1.1   Time-Sharing: "Successful" Research and Failed Products

Licklider faced a software crisis of his own around the time that he wrote his essay for Wiesner. As we saw in chapter 3, when Licklider served as the first director of ARPA's Information Processing Technologies Office, he defined and promoted "software" research in terms of improving human-machine interactions. For example, he funded MIT's efforts to develop a time-shared computer, Project MAC (for "Multi-Access Computer" or "Man-and-Computer"). As Atsushi Akera has argued, Project MAC gained support by aligning multiple institutional goals—those of research, service, and production.[5] But, by 1968, when Licklider became the Director of Project MAC, the alignment was falling apart.

Things started out optimistically enough. Under the direction of mathematics professor Robert Fano, Project MAC researchers built on the Compatible Time Sharing System (CTSS) developed by Fernando Corbato and his colleagues. In 1962, the CTSS provided real-time, remote access to MIT's computation center for up to three simultaneous users.[6] To expand access to the CTSS, Project MAC researchers transferred the system from the MIT computation center's IBM 7090 to an enhanced IBM 7094. Since the 7094 operating system was identical to that of its predecessor, the transfer was simple. The new computer arrived in October 1963, and CTSS was up and running within a month. By March 1964, MIT's system could serve 24 remote users at once.[7]

To achieve their ultimate goal of a university-wide computer service, Fano's group wanted a specialized computer, with an operating system optimized for time-sharing. They contracted with General Electric for the hardware, which in turn subcontracted with Bell Laboratories for the operating system. Both General Electric and Bell Laboratories believed it might become profitable to produce systems that could sell computer time on a large scale, and MIT's project—soon dubbed the Multiplexed Information and Computing Service (MULTICS)—seemed like a good way to get started.[8]

Thus researchers committed to developing a service for MIT while also demonstrating a potential commercial product. In 1964, based on experience expanding access to the CTSS, the Project MAC researchers were optimistic. They promised to make an operational system available to MIT by the end of 1965. But by March 1966, Fano could only report that the system would be useable by September 1967. The schedule continued to slip. In December 1968, worried that IPTO might cancel MULTICS, Licklider promised a "solid, reliable, useful" system "by the end of September 1969."[9] He sought "some chance of convincing ourselves and the world that the MULTICS project has already been successful," or of quickly advancing the project just enough to make it "obvious to many that the operating system, itself, is indeed successful."[10]

Thus, the "Multicians" (as they called themselves) discovered a more extensive research effort than they'd anticipated. They hoped that a computer upgrade might resolve some of their problems. But a new director of MIT's Computation Center canceled the upgrade in 1968, recognizing that it would require more research and development. He concluded that "as a *research* enterprise, M.I.T. can produce software on its own that is without parallel," but "research of this kind mixes poorly with computer-service objectives. . . ."[11] Similarly, General Electric decided to cancel commercial production of the prototype computer it was developing for MIT, realizing that "what we had on our hands was a research project and not a product."[12] Bell Labs also withdrew from MULTICS in April 1969.[13]

Ironically, time-sharing efforts such as MULTICS were designed to simplify the process of programming by making computers more interactive. And in the long run, they did. Dennis Ritchie, one of the last Bell Labs "holdouts" working on MULTICS, recalled that his group had "directly felt" a stake in the success of the project; they wanted "not just a good environment in which to do programming, but a system around which a fellowship could form."[14] Back at Bell Labs, they developed a simpler operating system for time-sharing, UNIX, which eventually was widely adopted. Today, over 40 years after its first demonstration, UNIX remains a very popular programming environment.

Despite its ultimate successes, time-sharing software failed to provide the services promised in the 1960s. Although MULTICS became available for general use in October 1969, and could support 55 simultaneous users by the end of 1971, the system still crashed an average of one or two times a day.[15] And while small-scale time-sharing services were commercially successful in the mid-1960s and 1970s, large-scale services were never profitable. As efforts to develop the complex software for time-sharing services faltered, a key assumption underlying time-sharing—that hardware was the most expensive part of computing—began to unravel.[16]

1.2   Marketing System 360: Revolutionary Product or Research Project?

At IBM, many computer experts anticipated that the time-sharing systems would be challenging to program. They sought a different way of cutting software costs: standardizing computer operating systems. In the 1950s and 1960s, computer manufacturers typically developed a unique operating system for every new computer model. And since manufacturers also typically bundled in programming support with each sale, the "software" became a significant part of the cost of both selling and running a new model of computer. For example, none of IBM's first six transistorized computers could run programs written for the other, requiring the company to support expensive programming for a wide range of machines. And though the dropping cost of processing power encouraged insurance, banking, and manufacturing industries to computerize, companies also resisted upgrading to newer and faster machines, dreading the expense of training employees to use completely different operating systems and programming languages.[17]

On April 7, 1964, IBM unveiled its solution: System/360. At a press conference in New York City, and simultaneous meetings in 165 cities around the United States, IBM announced: "The entire concept of computers has changed."[18] The "concept" included six computer models, each offering a different level of processing power and memory. The term "System" emphasized that these models would be unified by compatible software, while "360" underscored the system's versatility for business, control, scientific/engineering, and communications applications.[19]

Colorful brochures illustrated the system, with a single computer at the center of a room, surrounded by a ring of peripheral devices. Four very different potential computer users—an engineer, a businessman, a workman, and a woman (no doubt a secretary)—stood in the ring (figure 5.1). IBM promised: "Anyone who needs the services of your computer installation can write a program with a minimum of training. . . . It is this kind of programming support that can help cut your programming costs, reduce idle computer time, and keep your installation running at maximum efficiency."[20]

Figure 5.1
Operating System 360 advertisement, 1964. Reprint courtesy of Computer History Museum and IBM, © IBM.

Such was the vision in April 1964. And it sold, with over 1,000 orders for Operating System 360 (or OS/360) placed in the first month. IBM promised to roll out its smaller models by the summer of 1965, and its larger models in 1966.[21]

However, IBM was selling a product that had yet to be fully researched and developed. Within only a few months, the team designing the operating system discovered that it would not work well for the smallest model of the computer, because the memory was too small (only 32 kilobytes). The project was suddenly in dire straits. On the last day of 1964, IBM announced a new stop-gap plan: three operating systems designed to operate with different levels of memory, but all compatible.[22]

With this rapidly expanding programming job, IBM spent more money in the first year of software development than the company had projected for the entire effort. The most commonly used version of OS/360, the Disk Operating System (DOS), was finally delivered in June 1966. Over the next five years, IBM released some 26 upgrades of DOS in ongoing efforts to improve the system and eliminate glitches.[23] When IBM announced the "functional stabilization" of OS/360 in 1972, one bemused observer recognized it as "a corporate euphemism for euthanasia," and wrote an obituary:

The offspring first saw the light of day in December 1965 and the birth announcement recorded a weight of 64K... The huge weight of OS at birth contributed greatly to its early ill health. At the age of 3, OS began to show signs of long-term survival. ... OS was able to manifest itself in a bewildering variety of shapes and sizes. All of these were closely related and displayed the same genetic deficiencies, such as excessive size and poor health. Although self-reproduction was theoretically possible in 6 to 8 hours, most cases took 3 to 6 months to attain reasonable health.

... OS became increasingly complex, due mainly to the very large number of organ transplants. ... In its prime OS achieved a hitherto unprecedented size and, like the cowbird, kept growing until it filled its environment. This is believed to have been a significant factor in the growth of its corporate parent.

... It became apparent the OS had reached the level of its incompetence. Indeed, usually reliable sources indicate that it had already been promoted at least one level above that. OS ... will be mourned by its many friends and particularly by the over 10,000 system programmers throughout the world who owe their jobs to its existence.[24]

A few years after OS/360 was laid to rest, the manager of the programming effort, Frederick Brooks, published what became a much-celebrated text, *The Mythical Man Month* (1975). Brooks argued that a tremendously complex project such as OS/360 could not be hurried by adding more "manpower." Although the natural reaction to a delayed software project was to hire more programmers, like "dousing a fire with gasoline, this makes matters worse, much worse."[25] Since some parts of a project could not begin until others were completed, and new programmers required training, additional hires often diverted programmers from overdue steps in the process. "Simplifying outrageously," he coined Brooks' Law: "Adding manpower to a late software project makes it later."[26] Brooks argued that managers should instead adopt the stubborn attitude of a French chef: "Good cooking takes time. If you are made to wait, it is to serve you better, and to please you." [27]

1.3   A Software Crisis? The Conflation of Research, Development, and Production

Like time-sharing research, OS/360 was designed to simplify programming, and eventually did. OS/360 established operating system compatibility as an industry standard, allowing computer users to keep their programs when upgrading to a new computer. But these long-term successes entailed near-term failures in computer services—and embarrassment among would-be computer scientists.

For elite physicists, slow computation was worse than embarrassing; it was a dangerous weakness in the nation's scientific manpower. In early 1967, while serving as the U.S. delegate to the NATO Science Committee, the preeminent nuclear physicist Isidor Rabi expressed "deep concern about delays in delivering operating systems and other software" for large computers.[28] The German delegate to the committee consulted a colleague, Friedrich (Fritz) L. Bauer, from the Technical School in Munich. Bauer soon helped convene a study group, followed by a NATO-sponsored conference in "software engineering."[29]

In October 1968, 50 carefully selected computer experts gathered for the world's first "Software Engineering" conference, in Garmisch, Germany. Many spoke of a "software crisis." Edward E. David, the head of Bell Labs' part of Project MAC, and his colleague David Fraser, expressed alarm about "the seemingly unavoidable fallibility of large software, since a malfunction in an advanced hardware-software system can be a matter of life and death, not only for individuals, but also for vehicles carrying hundreds of people and ultimately for nations as well."[30]

Others objected to the language of "crisis." IBM's R. C. Hastings was "disturbed" by the "aura of gloom" because he was satisfied with his work on OS/360 installations: "These are complex systems, being used for many very sophisticated applications. People are doing what they need to do, at a much lower cost than ever before; and they seem to be reasonably satisfied."[31]

Brian Randell, an IBM researcher acknowledged "many good systems" but asked: "are any of these good enough to have human life tied on-line to them, in the sense that if they fail for more than a few seconds, there is a fair chance of one or more people being killed?" He feared "what might happen directly as a result of failure in an automated air traffic control system. . . . I am worried that our abilities as software designers and producers have been oversold."[32] Robert Graham, one of the Multics developers, went further: "an uncritical belief in the validity of computer-produced results . . . was at least a contributory cause of a faulty aircraft design that led to several serious air crashes." [33]

Kenneth Kolence who had recently cofounded a software consulting firm in Palo Alto, California, continued to object: "I do not like the use of the word crisis. It's a very emotional word."[34] He noted "many areas where there is no such thing as a crisis—sort routines, payroll applications, for example."[35] MIT's Douglas Ross countered: "It makes no difference if my legs, arms, brain and digestive tract are in fine working condition if I am at the moment suffering from a heart attack. I am still very much in a crisis."[36]

Whether they called it a "crisis" or not, participants in the conference agreed on two major problems. First, the time and labor to develop new software usually far exceeded initial cost estimates. Second, the performance of software nearly always fell short of initial promises. David and Fraser blamed these problems on "the unfortunate telescoping of research, development, and production of an operational version within a single project effort."[37]

Significantly, many conference participants blamed this "telescoping" on social pressures, not on any intrinsic feature of software. Kolence noted that "manufacturers are always under pressure from the users to give them something that works even if it is not complete."[38] Similarly, Professor John Buxton from the University of Warwick noted "deep confusion between producing a software system for research and producing one for practical use. . . . You are in fact continually embarking on research, yet your salesmen disguise this to the customer as being just a production job."[39]

Thus, the "crisis" emerged from conflicting representations of software—as research, service, and production. This mismatch was exacerbated in the most innovative software projects, such as OS/360 and time-sharing. Indeed, these projects drew more discussion than any other kind of software development.[40] Leaders in computing could only launch cutting-edge research by promising big results, but the results could be embarrassing.

Nonetheless, software researchers remained optimistic about the future of their field. Professor Stanley Gill, director of the Centre for Computing

and Automation at Imperial College London, argued that "research, development, and production" were "not clearly distinguished" because software engineering was still new.[41] H. R. Gillette of Control Data Corporation
expressed similar optimism. "We perhaps have more examples of bad large
systems than good, but we are a young industry and are learning how to
do better."[42] And Edsger Dijkstra, a Dutch computer scientist who was well
known for his views on programming, felt that the "admission . . . of the
software failure . . . is the most refreshing experience I have had in a number of years, because the admission of shortcomings is the primary condition for improvement."[43]

2   Computer Experts Size Up Missile Defense

2.1   Underestimates and Overexpectations
Ten years after the first software engineering conference, Randell recalled
the missile defense debate and his "horror and incredulity that some computer people really believed that one could depend on massively complex
hardware and software systems to detonate one or more H-bombs at exactly
the right time and place over New York City to destroy just the incoming
missiles, rather than the city and its inhabitants."[44] Despite these recollections, it seems that missile defense received no mention at the first software
engineering conference.[45]

Instead, arguments about missile defense software first emerged at the
prompting of elite physicists who were working with policymakers. In February 1969, Senator Edward Kennedy asked two prominent scholars and
former advisors to President Kennedy—Jerome Wiesner and Harvard Law
professor Abram Chayes—to pull together a study of missile defense.[46]
While Chayes focused on political and diplomatic challenges of defense,
Wiesner worked with Steven Weinberg, a theoretical physicist at MIT, to
gather essays on technical issues.

Wiesner and Weinberg started with a long list of technical issues and
associated experts, putting "reliability" at the top. Their list of experts on
reliability included several physicists in the arms control community who
had also managed the development of complex technology, including
Jack Ruina, Herbert York, and three physicists associated with the Stanford
Linear Accelerator—MIT professor Henry Kendall, and Stanford professors
Wolfgang Panofsky and Sidney Drell.[47]

Some of these physicists called attention to the unreliability of complex computers. Kendall noted "widespread experience that in large computer installations, even after weeks of apparently satisfactory operation,

the entire system can fail from an entirely unsuspected demand on the program, unanticipated in the software design."[48] He suggested that the "subtle mingling of software and hardware" in missile defense would be untrustworthy without "live tests"—meaning atmospheric nuclear detonations, which were forbidden.[49] He continued: "There is no way to check a computer even if it is asked to do a limited job much less carry out a complete stratospheric war."[50]

Wiesner soon reached out to friends involved in computer development. At Licklider's suggestion, Wiesner contacted Edward E. David, who had recently given a talk on the challenges of developing software for the telephone network's Electronic Switching System (ESS).[51] Though he had spoken at the software engineering conference of "unavoidable" software glitches that could be a "matter of life and death," David was committed to deploying defense. He explained:

> . . . Fail-safe operation cannot be assured 100% but careful testing, design, and failure protection features can reduce the probability of a catastrophe to some small but as yet indeterminate number. This uncertainty arises because software failure mechanisms are not understood in a sense which permits statistical estimation of their frequency. When one thinks of this situation in an "ultimate" context it is a bit unnerving but certainly not unique since human failure has the same unpredictability about it.[52]

David, who worked for the Safeguard contractor (Bell Labs) and who would soon become President Nixon's science advisor, seemed committed to deploying missile defense, if only because he saw no better alternative.

By contrast, Licklider was eager to critique the missile defense system. Even though Project MAC depended on Defense Department funding, many of the project members opposed the Vietnam War and what they viewed as wasteful military spending. Thus, Licklider easily found sympathetic professors who were eager to highlight the risks of missile defense computers.

Nonetheless, his colleagues also expressed "skepticism about the validity of reasoning from available evidence" to missile defense.[53] After a couple of weeks spent gathering information, Licklider remained disappointed about finding only "quasi-facts."[54] Part of the problem was that software developers had little interest in publicizing failures. But more fundamentally, Licklider suggested that a list of failures simply would not yield a "truly authoritative" analysis.[55] The best he could offer was "a subjective appreciation" of the risks associated with developing complex systems.[56]

In this spirit, Licklider described a class of systems that were characterized by high complexity, unpredictable and changing operating conditions, and "difficulties in the way of testing."[57] Prominent examples included SAGE, the Strategic Air Command Control System, the SABRE airline reservation system and OS/360. All of these systems, he emphasized, took longer to develop and cost more than initially expected.

If missile defense software could help protect populations, high costs and slow development might be acceptable. But Licklider insisted that the system would not be trustworthy. He explained that complex software could only be "'mastered' (which is not the same as 'perfected')" if it were "developed progressively, with the aid of extensive testing," and then "operated more or less continually in a somewhat lenient and forgiving environment." By contrast, since missile defense hardware and software would face a constantly changing threat, "the whole system effort would be reduced to the software's pace and to its state of confusion."[58] He concluded: "The sad plight of the software system might be hidden . . . until the arrival of the hoped-against moment of truth. Then would the bugs come out."[59]

2.2   Computer Professionals Against the ABM

While Licklider's report was commissioned by well-connected physicists, the "Computer Professionals Against the ABM" emerged as more of a grassroots movement. The group was initiated by Daniel D. McCracken, a writer and consultant whose books on programming became canonical in the 1960s. McCracken earned a bachelor's degree in chemistry before turning to computing, where he wrote not only about programming techniques, but also about the professional responsibility of programmers. By the late 1960s, he was a speaker on the ACM national lecture circuit, an antiwar activist and a student at Union Theological Seminary in New York.[60]

Near the end of May 1969, McCracken wrote to several friends who might be interested in "forming a committee of computer people to oppose" missile defense.[61] They included Paul Armer at Stanford's computing center, Joe Weizenbaum at MIT, and Gregory Williams at General Electric. McCracken acknowledged that he opposed Safeguard as an escalation of the arms race and misdirection of national resources, but emphasized a technical problem: since the software would be complex, large, and "essentially untestable," it would likely be unreliable.[62] He proposed that they write a report about the software problem, and develop a statement that could be signed by prominent computer experts.

McCracken's missive hit its mark. Weizenbaum, Williams, and Armer agreed to form an executive committee for the movement. In mid-June 1969, Weizenbaum and McCracken spent a day in Boston, forging a strategy. Weizenbaum urged that the group "stick 100% to the computing feasibility issue" because it was "what we know most about." McCracken agreed, and soon sent a draft statement to the "exec."

We, the undersigned members of the computing profession, wish to record our professional judgment that there are grave doubts as to the technical feasibility of the computer portion of the Safeguard Antiballistic Missile system. These doubts range from a profound skepticism that the computing system could be made to work, to a conviction that it could not.[63]

The statement drew analogies between missile defense and other complex software systems, arguing that without operational testing, the defensive system would never be trustworthy.

The group was astute to avoid political issues, as the ACM and other professional societies were torn by controversy over their appropriate role in issues such as the Vietnam War. In fact, several members of the ACM had recently called for the association to issue a statement against the Vietnam War. Although this statement fell outside the constitutional charter of the ACM, it raised a provocative question: "should the constitution of the ACM be revised to permit comment on deeply social or political issues?" The question was referred to the ACM membership for a general vote in April 1969.[64] However, it was resoundingly defeated, and ACM president Bernard Galler was "amazed at the large number of letters" protesting that the ACM had even taken a vote.[65] In his August 1969 letter to members, Galler encouraged members to get involved in political issues "personally, but not through their professional association."[66]

Thus, McCracken stirred considerable controversy when—before finalizing a statement with the "executive committee"—he wrote to about 200 people, inviting them to oppose defensive deployments as "both bad national policy and bad computing."[67] McCracken explained that since "computers are at the very core" of missile defenses, the ACM could easily address the technical feasibility of the system within the existing constitutional charter.[68] In a letter sent to all ACM councilors, he argued that missile defense was an example of "the sort of issue that . . . the ACM should be involved in as a body."[69]

Even council members who agreed that the defensive computers were technically dubious were wary of involving the association in the missile defense debate, because they questioned whether they had a sufficiently "technical" basis for opposition. In August, Galler wrote to McCracken, explaining that he "essentially" agreed that the missile defense software would not likely be reliable, and had written "as strong a letter as I could" to his senator.[70] But Galler would not sign the statement against the ABM because it made too many claims that he "could not defend if asked."[71]

Other leaders of the ACM joined or even sponsored the Computer Professionals Against the ABM, but refused to involve the association. For example, Anthony Ralston, a professor at the State University of New York Buffalo who would soon become the ACM president, wrote:

I am entirely on your side not only because of doubts about the computer system feasibility, but also, more generally, about the rest of the system. More years ago than I care to think about I spent two of the least productive years of my life working on Nike Zeus at Bell Labs.[72]

While Ralston endorsed the statement, he also felt that "if professional societies like ACM take stands on such issues, they will tear themselves apart."[73]

2.3   Experts and Expertise in Dispute

What the Computer Professionals Against the ABM lacked in formal professional backing, they sought to gain in the form of elite "sponsors," who would add their names to the CPAABM letterhead. They targeted experts associated with well-known computing innovations or companies, including the occasional physicist—such as Richard Garwin, who retained a long-time affiliation with IBM's Watson Research Laboratory. Significantly, only three of forty targeted sponsors were also participants in the first software engineering conference, and only one of these—Robert Bemer, who was also a friend of Armer's—signed on as a sponsor.[74] "Software engineering" was barely visible to the computing community at large.[75]

Nonetheless, the exec quickly gained nearly two-dozen sponsors, including Ralston, several of Wiezenbaum' colleagues at Project MAC—Corbato, Fano, Licklider, and Minsky—and Armer's colleagues in Stanford's artificial intelligence group—Edward Feigenbaum, Donald Knuth, and John McCarthy. Eventually they gained the sponsorship of FORTRAN "inventor" John Backus, and Fred Brooks.[76]

In an effort to gain as much publicity and as many signatures as possible, McCracken contacted his friends in the trade press. He started with Robert Forest, an editor at *Datamation*, a magazine popular for "its irreverent view of goings-on in the industry."[77] He also contacted the quirky editor and founder of *Computers and Automation*, Edmund Berkeley.[78] He asked them to publish a letter inviting readers to oppose missile defense deployments because the computer systems were technically dubious. Berkeley and Forest both enthusiastically publicized the movement.[79]

With about 200 signatures by the end of June 1969, the executive held a press conference, and over the course of the summer, they sent their statement to over 600 contacts, inviting them to join "The Computer Professionals Against the ABM."[80]

But what did it mean to be a "computer professional"? McCracken named the group with ambivalence. He acknowledged that the term "'professional' has its problems," but preferred it to "scientist," noting: "I myself have never appropriated that title." [81]

These comments reflected broader ambiguities surrounding computing expertise in the late 1960s. As Thomas Haigh has discussed, computers opened up space for many different kinds of work, and therefore many different kinds of professionalization projects.[82] For example, "systems men" promoted computers as a way to revolutionize corporate administration. Separately, managers of data processing units sought to elevate their professional status as they began overseeing rooms full of highly modern electronic computers, rather than tabulating machines. Both systems men and data processing managers emphasized their superiority to programmers, whom they portrayed as mere "technicians." But programmers were in such high demand, and their skills so poorly defined, that they gained many of the privileges of "professionals."[83]

For their part, the academics that dominated the ACM were struggling to gain legitimacy as "computer scientists," and did not want to be training workers for industry.[84] In 1966, when a young University of Maryland professor, David Parnas, weighed in on the ACM curriculum recommendations, he emphasized that mere technical skills (such as programming) were less important than the ability to "innovate and work on previously unsolved problems"—that is, to conduct research.[85] Computer scientists sought to gain academic legitimacy by underscoring theory. By 1969, when 25 top universities were granting PhDs in variants of "computer science," the algorithm was becoming the basis for a highly theoretical university curriculum.[86]

As Nathan Ensmenger has discussed, academic approaches to training often clashed with industry's emphasis on experience and certification exams. Ultimately, employers would not hire on the basis of academic degrees, professional certification, or even membership in a professional society. By 1968, the *Wall Street Journal* reported that less than 13 percent of programmers belonged to a professional society.[87]

Without a well-defined basis for computing expertise, many computer workers questioned their qualifications to join the Computer Professionals Against the ABM. Most of those who signed on as "computer professionals" grounded their qualifications in specific working experience rather than credentials or formal training. One wrote: "I am not sure as to the implications of the use of the word 'professional' in the committee's title, but if you have set up admission requirements feel free to request a resume."[88] Others described experience in the defense industry:

... I have just joined the research group at Lockheed after spending eight years at MIT. I was associated with Project MAC for four years. ... I spent one summer working for the Information Processing Techniques Office of the Advanced Research Projects Agency.

... For the past two years I have been a member of the Defense Science Board Task Force on Computer Security. I have even heard classified briefings on the computer systems for Nike-X, the predecessor to ABM. All of my experience just strengthens my conviction that ABM is technically infeasible.[89]

Several respondents emphasized similar experience in defense or aerospace computing, focusing on specific companies and projects.[90]

By the same token, many workers who had no specific experience with military computing felt that they had little technical basis for opposing missile defense. For example, one wrote: "I consider myself a computer professional; however, working entirely in the commercial area, I probably don't have grounds for opposing ABM other than ethical."[91]

Indeed, several individuals charged that the computer professionals' movement could not speak authoritatively without inside information. Willis Ware, who studied computer security at RAND, asked: "can't your position in turn be challenged on the grounds that you, from an unclassified position, really don't know any details of the computing job involved?"[92] McCracken replied by pointing to "private communications" indicating that "technical people at Bell Labs have no idea how to solve some of the problems."[93] He also tried to arrange a public debate with an insider, targeting individuals such as Richard Hamming, a prominent mathematician and signal processing expert at Bell Labs. But Hamming would not condescend to a debate with McCracken, and discounted his "private communications":

... I suspect that you talked to programmers who per usual have a myopic view and often don't know what they are talking about. I privately made my own survey and I found *no* one who seriously doubted that we could and would deliver the goods according to the specifications as laid down in a thick, secret document. ...

Since you (and your friends) do not and cannot know what is in the specifications, how can you have a serious opinion on the matter?[94]

McCracken retorted: "If an inventor came to you with a thick book of specifications for a perpetual motion machine, would you need to read it?"[95]

But there was the rub: computer experts had no equivalent to the physicists' second law of thermodynamics. At the same time, we must ask: why did computer experts need to prove the impossibility of developing reliable software? Why wasn't inductive analysis sufficient proof?

2.4   Sufficient Proof? Professional Responsibility and Technological Progress

Political commitments likely drove some computer experts to insist that the software was feasible, unless it could be proven otherwise. Many were deeply involved with Defense Department computing. For example, Roy Nutt, the cofounder of Computer Sciences Corporation, a major software developer with both civilian and defense contracts, rejected McCracken's invitation to join the CPAABM: "As a member of the National Academy of Sciences Review Committee for Safeguard Data Processing, I must refuse any association such as you mention." He continued heatedly:

First, while the computer system is of course very complicated, it is not out of reach.

Second, how many megadeaths must you prevent to justify ABM? Nuclear war is a terrible thing to comprehend . . . but in that frightful event we must somehow guarantee survival. . . . Remember the ostrich!"[96]

McCracken objected that "Safeguard is not designed to save megadeaths, it is designed to save a few dozen missiles so that we could cause even more megadeaths after a first strike." He concluded: "Remember the lemming!"[97]

At a time when activists were taking aim at the "military-industrial complex," McCracken tended to attribute the views of individuals like Nutt to conflicts of interest. In early 1970, he found an opportunity to test his hypothesis. Anthony Oettinger, the chairman of the National Academy of Science's Computer Science and Engineering Board, invited him to an upcoming meeting.[98] There, McCracken immediately began digging for information on the Safeguard Data Processing committee mentioned by Nutt. He made contact with the committee's chairman, Professor Bruce Arden at the University of Michigan, and from there obtained names of all the members. He determined that 8 of 13 committee members worked for defense contractors with Safeguard contracts. Although Arden pointed out that none of the individuals on the panel were actually engaged in missile defense work, McCracken remained skeptical.[99] He wrote a story for the editor at *Computerworld*, Joe Hanlon, who on August 5 eagerly published a front-page article headlined: "ABM Computer Advisors in Interest Conflict?"[100]

But while some computer experts may have insisted that missile defense software was within reach because of *political* or *financial* commitments, many others refused to join the CPAABM because of a far deeper *vocational* commitment to technical progress. Herbert Bright, an ACM council member, told the leaders of the computer professionals' movement that he was "dubious" about their basis for speaking out:

Philosophically, it's tough to prove that something can't be done unless it would violate a known physical or mathematical law. . . .

Although I feel, gutwise, that you're right in considering the system unpredictable, I also feel it's unfeasible to develop a sound basis for a proof that the proposed system cannot work; and I have a profound distaste for saying we believe it won't work when the truth is merely that we suspect it won't work. Let's not be intellectually dishonest with the public in implying that we know what we only suspect. . . [101]

Weizenbaum replied that he felt "somewhat misunderstood."

I don't believe any of us are in a position to "prove" that the computational subsystem of the ABM system won't work. I personally suspect it won't work. My suspicion is strong to the point of being belief. I don't think that my statement as a professional that I hold this belief obligates me to a mathematical proof.[102]

Bright ceded this point. But he insisted that by the time defenses were deployed, "We computerniks might learn how to debug such a complicated system (of course, we don't know how now), and . . . new technologies in sensors and/or weapons might cure the weaknesses we now see. . . ."[103] Weizenbaum countered this optimism by highlighting the "fundamental reasoning . . . that large computing systems are products of evolutionary development."[104] He argued that large computer systems only became reliable through a process of slow testing and adaptation to an operational environment. Computer systems could only become reliable if "the environment that they are to control or with which they are intended to cooperate must have a change rate smaller than that of the [computer] system itself."[105] But since the missile defense "environment is actively hostile and destabilizing by intent," the software could not keep up with its changing requirements: "convergence is logically impossible." Weizenbaum concluded: "This is not a question of not knowing enough technology *now*. There is a difficulty *in principle* here."[106]

Wiezenbaum's arguments, couched in the mathematical language of convergence and logic, left Bright feeling "better informed, more worried, and less decisive" about ABM deployment. He replied: "Good luck in your fight!"[107] Yet Bright did not join the movement.

Even computer workers who opposed missile defense for political reasons were optimistic about technological progress. One wrote: "Given the time and money Yankee ingenuity could probably make the damn thing work but I'm opposed to such waste when people are hungry, not to mention the stupidity of arms escalation."[108] Another was "not wholly convinced that human ingenuity isn't up to the task of building the system," even if at a "prohibitively high cost," but felt that technical arguments were "needed to convince our legislators."[109]

Gerald Salton, professor of computer science at Cornell, expressed his "sympathy with efforts" to stop deployment of missile defense, but would not join them:

I simply do not believe that based on the evidence which you submit any computer expert could certify that the software problems connected with the ABM system are really beyond our reach. . . . If any real data exists about the details of the software required by the ABM system . . . I would be most interested in reconsidering this most important question.[110]

Many computer experts felt a vocational commitment to working for progress even in the face of doubt. One wrote: "The astounding tasks of recent years have been accomplished only because people had the courage and fortitude to believe they could be done, even though a path was not apparent. . . . Surely the computer profession does not consist of people with a 'give-up' nature."[111] Similarly, Robert Head, who had helped manage programming for the SABRE airline reservations system, wrote:

I am not close enough to the technical aspects of the question to be able to make intelligent judgments, though I believe that there is an analogy between this effort and our man-in-space effort in the sense that both are very ambitious technological undertakings. . . . I am hopeful that the computer support for the ABM system will prove as effective as that which has been forthcoming in the Mercury, Gemini, and Apollo space missions.[112]

Thus technical progress was tied not only to *professional* identity, but to *national* identity in the era of Apollo. Computerniks argued that "Apollo 11 is just as implausible as ABM," and "How certain was Apollo in 1961?"[113]

These sentiments were remarkably prevalent in responses to the CPAABM. McCracken's initial letters prompted replies from about 111 respondents, but only about half of these (57) wrote to join the movement. Nearly one-third of the letters (34) came from individuals who refused to sign on, while the rest (20) were noncommittal. Perhaps most strikingly, 17 individuals went out of their way to argue that the software could be developed.[114] Confidence in technological progress undermined efforts to gain consensus among computer experts.

3   Computing, Risk, and Progress in the Safeguard Debate

3.1   Computing and Common Sense

Although computer experts did not demonstrate consensus about the risks of complex software, their arguments were eagerly embraced by Safeguard critics. When McCracken wrote to all members of the Senate in early June, explaining that he was organizing a group of computer professionals to argue that a missile defense system was technically infeasible, Safeguard opponents were immediately interested. Fulbright, Hart, and Gore each requested a statement for the Congressional Record. Gore explained that his committee was "considering holding a specific hearing on computer problems."[115] Hart viewed the movement as "one more piece of ammunition supporting the argument that Safeguard is just unlikely to work as advertised."[116] By mid-July, with approximately 300 signatures gathered, McCracken's House representative inserted the Computer Professionals' statement into the Congressional Record.[117]

The CPAABM was also embraced by the "National Citizens Committee Concerned about Deployment of the ABM"—a group that formed in late March 1969 with the sponsorship of Roswell Gilpatric, Gerard Piel, Herbert York, and other leaders.[118] The "Citizen's Committee" aimed to focus several grassroots efforts into a unified opposition to Safeguard. In late June it helped Weizenbaum, McCracken, and Armer hold their press conference.[119]

Nonetheless, physicists continued to dominate public debate, even when questions about computing emerged. In early May, preparing for a second round of hearings, Senator Kennedy began circulating Wiesner and Chayes's edited report to other members of Congress and to the press. Spurred by Licklider's essay, "Underestimates and Overexpectations," Wiesner devoted considerable discussion to computer reliability when he testified on May 14. He emphasized that software was "more difficult than" hardware, that very large "computer programs can only be tested properly in their operational environment," and that Safeguard could not be so tested—leaving it prone to failure.[120]

Arguments about software glitches fared well in public because they had a commonsense appeal. Glitch-ridden computers became a favorite talking point for senators opposed to Safeguard. Fulbright questioned Donald Brennan, who favored defenses as a form of arms control and of reducing U.S. fatalities in the event of war:

When I read nearly every day or so of some strange aberration of our computers, I . . . have grave doubts that this system will work any better than the computers which have been under development by IBM for years and cost a million or $5 million each. They don't work. They chew up checks. They give people wrong checks and do all sorts of strange things. Why do you think they are going to be perfect if they control a Sprint, Dr. Brennan? Will they? Does this give it some special sanctification?[121]

Brennan replied: "It isn't the Sprint that gives them the special sanctification, it is the amount of money spent on it."[122] He claimed that weapons systems were "made on a completely different basis than the kind of commercial systems in which some failures are tolerable," and that "at least some" such computer systems "do not fail because of simple failures of electronics. . . ."[123]

In suggesting that software reliability could be achieved through increased funding, Brennan treated software as a "black-boxed" product, rather than as a process of evolutionary research and development. Similarly, Deputy Secretary of Defense David Packard acknowledged that the missile defense "data processing job is a large one."[124] But he continued: "It does not involve any new technology. It is simply a large system involving data processing."[125] He explained that "the availability . . . of the skilled people doing that part of the job is an important factor in the effective operation of the system."[126] And having studied "the various aspects of the system's components," including software, Packard concluded that "the system does have the capability to do the job that we expect it to do."[127]

Likewise, John Foster, director of defense research and engineering, argued that it would be possible to develop a "computer to handle the vast amount of data necessary" for defense. He critiqued the Chayes-Wiesner report, including its analysis of computers, arguing: "A mammoth computer currently in operation and similar in design to what will be used with Safeguard contains almost one million different instructions."[128]

Foster's comments were not necessarily persuasive. The *New York Times* editorial staff objected that Foster did not address the difficulty of "overcoming programming errors," a major issue in the Chayes-Wiesner report. But they continued: "The main argument against Safeguard lies elsewhere. Even if an effective antiballistic-missile system could be built . . . there is no need for it at this time . . . it would probably start a new round in the arms race and . . . it 'would seriously impede' agreement in the impending strategic arms limitations talks with the Soviet Union."[129]

3.2   Complex Technology, National Security, and Testing

Indeed, by the spring of 1969, physicists increasingly tied arguments about fallible software to a broader debate about the best way to enhance national security. Would it be better to rely on missile defenses, even though their tremendously complex hardware-software systems could not be operationally tested, or to forego defenses and instead negotiate offensive reductions?

As we have seen, York was a leader of the many physicists who came to prefer negotiated limits. From the early 1960s onward, York argued that a complex defensive system could not possibly be tested against all of the surprises an attacker might present. With only a "forlorn hope" of defense, he argued that complex technology could not solve the problems of national security.[130]

But other physicists put more confidence in complex technology than negotiations, and this shaped their views of what constituted sufficiently "realistic" testing. When York asked his longtime friend and colleague "Johnny" Foster to address the likelihood that defenses could cope with surprise attack, Foster argued that an "out of the blue" attack "seems quite unlikely."[131] Instead: "The most likely attack (ICBMs from Russia) will almost certainly be detected very shortly after launch . . . so the question then is, with tactical warning (and probably some strategic warning), will the system be able to react?"[132]

Remarkably, Foster's anxiety in the 1963 test ban debate—that missile defense systems were too complex to be proven without atmospheric nuclear testing—had dissipated. He argued: "We can test all of the elements of the tactical system with the exception of the warhead itself. Here the combination of underground testing and the meticulous design practice of the AEC [Atomic Energy Commission] gives us the highest confidence of successful operation in time of nuclear attack."[133]

The differences between York and Foster illustrate the subtle relationship between commitments to technology and views of "realistic" testing. As Trevor Pinch and Donald MacKenzie have emphasized, notions of "realistic" testing are socially shaped.[134] The missile defense debate suggests that views of realistic testing also influenced political commitments. York preferred negotiated agreements in part because he believed that tremendously complex missile defense systems could not be fully tested against all of the attacker's surprises. By contrast, Foster discounted the likelihood of a surprise attack, preferring even a partially tested defense to an arms control treaty.

Foster was so confident of the potential value of defense that he took the time to write a rebuttal of the CPAABM arguments.

There is nothing about the system that essentially hasn't been done before. Computer-controlled phased array radars are operating in the field today. Hundreds of missiles have been fired, most under computer count-down control. Many antiballistic missiles have been fired and successfully tracked and guided to intercepts of live targets. Computers are being used to monitor all aspects of very complicated systems, determine faults and failures, and immediately reconfigure the system accordingly. . . .

Thus, the only real tasks that the Safeguard system has is to integrate all of these functions in the computer programs and to check thoroughly and test out the programs before the system is made operational.[135]

In an argument published in *Modern Data*, Foster continued by describing plans to test the programs with simulations in a dedicated software development facility.

To many computer experts, Foster's remarks reflected blithe ignorance of the vast difference between a developmental software system, and one that would be trusted as a finished product. McCracken used these remarks for an "essay contest." In the pages of *Computers and Automation*, he invited readers to address the question, "Would you trust the lives of your children to OS/360?"[136] This line proved to be a reliable source of laughs when McCracken began traveling to ACM chapters around the nation to speak on "Why the ABM computer system won't work."[137]

4   Conclusion: Black-Boxing Software in the Age of Apollo

In sum: although computer experts and a few physicists highlighted the risks of fallible computers, expert consensus on the challenges of missile defense software was nowhere to be found in the late 1960s. And without consensus, many policymakers were inclined to place their faith in technological progress.

In July 1969, the majority of the Senate Armed Services Committee recommended that the first phase of the Safeguard deployment be funded. Their arguments focused on the growing threat from the Soviet Union and the need to maintain a strong negotiating posture. They noted that some "scientists have testified that the system will not operate effectively due to its complexity while other scientists have testified that . . . the technical problems can be resolved." They concluded: "It is prudent that any doubt in this question be resolved in favor of confidence in the system."[138]

By contrast, a minority of the committee opposed deploying Safeguard. This group doubted that the radars, which had yet to be fully built and tested, would "operate together in that almost instantaneous manner which would be necessary in case of sudden attack."[139] They were even more dubious "that the computer, which has been neither built nor tested, and which is admittedly far more complicated than any computer ever yet attempted, will operate properly when called upon to do so."[140]

When the full Senate considered these conflicting recommendations on August 6, 1969, it was faith in technological progress that prevailed. It was the spirit of the times. The final Senate debate took place just two weeks after Buzz Aldrin and Neil Armstrong planted an American flag on the moon. In closing remarks, Senator Henry Jackson declared: "We won't settle an issue like the ABM by claiming it won't work. . . . We solved greater and more complicated technical problems to make the successful Apollo flight than we found in the Safeguard program. Surely, if we can walk on the moon, we can make the Safeguard program work."[141] The very same evening, in a 51–50 vote with Vice President Spiro Agnew casting the tie-breaking vote, the Senate narrowly approved Nixon's request to begin construction on the first two Safeguard sites, in Montana and North Dakota.[142]

Neither the close Senate vote, nor the House's easy approval of the Safeguard "Phase I" deployment in October of 1969, ended the debate. As deployment began, and as the Strategic Arms Limitations Talks continued in the early 1970s, scientists and policymakers continued to argue about what defensive technology the United States should be deploying, if any.

As we will see in the coming chapters, software would be integral to this debate, in no small part because the question remained: was Safeguard a research program, a developmental system, or a weapon in production? Many computer experts persistently argued that without operational experience, missile defense software could never be a trustworthy product. The Computer Professionals Against the ABM, represented primarily by McCracken, continued to work with physicists in the arms control community, and to generate debate among computer experts.[143] Several physicists and policymakers took arguments about fallible computers into the mainstream Safeguard debate. But without any disciplinary repertoire, neither physicists nor computer experts could persuasively demonstrate the limits of software reliability. Expressing twin ideologies of nationalism and technological enthusiasm, policymakers brushed aside doubts about software, and continued to vote Safeguard forward.

Nonetheless, the debate about missile defense software marked a turning point. The leaders of CPAABM—Dan McCracken, Paul Armer, and Joseph Weizenbaum—were among those who continued to press the ACM to engage with political and social issues. Although many of these efforts fizzled, in 1973, McCracken and Armer became leaders of a new ACM Committee on Public Policy.[144] As we will see in chapter 8, this committee would last, eventually becoming a locus for discussing the risks of complex software—including missile defense.

More visibly, elite physicists began to tailor their recommendations for missile defense around a growing appreciation for the challenges of complex software development. As the next two chapters show, the Defense Department responded to these critiques with a new research and development program that would slowly contribute to the rise of "software engineering." By any measure, the debate was a game-changer.

6   The Politics of Complex Technology

Bell Labs, the principal contractor for the Safeguard system, has apparently decided that Safeguard is not worth building.
—Lawrence Lynn to Henry Kissinger, April 14, 1970

In January 1970, as the Army prepared to deploy defenses around missile silos in North Dakota and Montana, President Nixon urged that this "Safeguard" system be expanded to defend cities and the nation's capital. He asserted that Safeguard would provide cities with a "virtually infallible" defense from a small missile attack.[1] But within a few months, he got bad news. Lawrence Lynn, the aide to Nixon's national security advisor, Henry Kissinger, warned that Bell Labs was unwilling "to be associated with a program which cannot technically perform the missions the Government claims it will accomplish."[2] Bell objected that Nixon kept changing Safeguard's mission, and worse, that the "system is being advertised as capable of doing things it was never designed to do, in any version. (That is, 'virtual infallibility.')"[3]

Lynn described Bell's explanation as a "not very exciting recital" of public critiques already made by physicists. Nixon was changing the defensive mission with the political winds, forcing Bell to constantly reengineer the technological system. Since the defenses were too complex to be rapidly redesigned, the deployment pace was slowing. By the time defenses were deployed, they would "face a later, more sophisticated threat" that rendered it obsolete.[4] Significantly, both Bell Labs and Nixon's science advisors noted that the "pacing factor" slowing any deployment was complex software.[5]

Nixon and his advisors didn't buy it. Lynn noted Bell's "highly self-serving" argument: "the components were designed for one mission and then the politicians changed the mission . . . everyone is at fault except

Ma Bell and she, conscience-stricken, won't have any more of it."[6] Nixon guessed that the "real reasons are their scientists and P.R. [public relations] fears."[7] Though the administration feared a "major embarrassment," it was not persuaded of technical problems.[8] Nixon and Kissinger continued to press for missile defenses, treating the system's mission as a flexible negotiating tool.

How, then, did the challenges of complex technology—and especially software—influence missile defense? This chapter shows how the complexities of defensive hardware and software shaped the most visible outcome of the late 1960s debate—an Anti-Ballistic Missile Treaty limiting defensive deployments—as well as the long-term missile defense program. Some scholars have analyzed the ABM treaty as a rational result of a technological reality—the tremendous physical advantage of offense over defense—while others have discussed it as a product of socially constructed ideas about arms control.[9] Yet, as we will see, neither explanation quite accounts for the twists and turns in the negotiations. Delegates recall anything but a rational, technology-driven process.[10] Documents declassified by William Burr and the National Security Archive underscore the ways that Kissinger micromanaged the proceedings, ignored scientists' advice, and undermined his own delegates.[11] At the same time, the treaty reflected far more than ideas—the technical details mattered.

In the late 1960s and early 1970s, physicists grew increasingly worried that strategic weapons were coming to rely upon complex, relatively inflexible computer systems. Though Nixon and Kissinger sought to use new weapons as flexible tools—for negotiating with the Russians, as well as in nuclear operations—they discovered that the complex technologies and interests surrounding missile defense undermined any such flexibility. The practical challenges of managing complex hardware and software both shaped, and were shaped by, missile defense politics, leading to the negotiation of the ABM treaty in 1972.

Yet the treaty was not necessarily the most important outcome of the debate. As we will see, the debate about missile defense and its software contributed to a broader transformation in the relationship between and among physicists, computer experts, and policymaking. As elite physicists participated in the public debate, they established new modes of political engagement, even as they undermined their inside connections. Physicists' arguments encouraged the missile defense program to prioritize software research. By the early 1970s, computing and software research would grow dramatically, while physics would find its privileged place challenged.

1   The Technopolitics of Defense

1.1   Discovering the Inflexibility of Software

In early 1969, with uncertain Congressional support, and impending Strategic Arms Limitations Talks (SALT), the Nixon administration sought maximum flexibility in the missile defense program. Yet Nixon's science advisors increasingly warned that defenses would not be so flexible—in no small part because they relied heavily on computers.

PSAC took its cues not from the computer experts who publicly opposed missile defenses, but from Bell Labs and other consultants to the missile defense project. Bell Labs had long been developing a specialized parallel processor to track missile defense targets, but the software required for this scheme was growing very complex. In April 1969, a National Academy of Sciences panel argued that commercial systems could better keep pace with the Soviet threat as it grew over time. The DDR&E, John Foster, was persuaded that commercial computers would eventually be best, but feared that a new approach would delay the deployment schedule.[12]

As the physicists on PSAC's Strategic Military Panel learned more about the software challenges, they concluded that Foster's approach was dead wrong. In December 1969, the panel chair, Sidney Drell, reported on the group's review of the Army's missile defense program: "Software developments are the pacing factors determining the date of initial deployment."[13] As a result of this slow pace, Safeguard could not make a significant contribution to protecting Minuteman silos before 1976. It might be useful in the late 1970s, but it would also become very expensive due to the growing Soviet arsenals. Noting that "rapid strides in the development of the computer industry" could reduce costs and speed deployment, the panel recommended deferring deployment, while continuing to study a wide range of options for hard site defense.[14]

Instead, the Nixon administration pressed on with plans to deploy missile defenses as soon as possible, using the same components that had been optimized for defending cities. Though Nixon initially announced that Safeguard would defend missile silos, hoping to ease local opposition to nuclear-tipped interceptors, he privately hoped for city defenses, and in January 1970 expanded the goals of the defense to include cities and the District of Columbia.[15]

Thus, physicists once again went public with their opposition to Nixon's plan. Panofsky argued that because the Nixon administration was trying to use the same components to defend two different kinds of targets—cities and missiles—the performance of both kinds of defense would suffer. Computers were just part of the problem. Panofsky noted that a "hard-point" defense would be cheaper and more effective if it used simpler radars and missiles than those required for city defense. A city defense required very costly radars and fast interceptors because it needed to kill missiles at high altitudes, and thereby minimize damage to the "soft" targets below. By contrast, Minuteman defenses needed only simple radars and interceptors because they could intercept missiles at lower altitudes, relatively close to hard missile silos. Additionally, radars for silo defenses required hardening from nuclear blast, whereas there was no need to harden radars for city targets.[16]

Additionally, Panofsky noted that the deployment schedule of any missile defense would be "paced by the unprecedented complexity of the computer and the associated programming," and hence could not be useful for several years.[17] If "small, autonomous radars" rather than the more complex radars were deployed around silos, the software could be developed more quickly.[18] Furthermore, he argued that "the performance of the system is limited . . . against sophisticated attacks by anticipated software capability."[19] Panofsky and like-minded physicists warned that by deploying the same components for different kinds of defense, the administration was committing to a white elephant, not a flexible weapon.

The problem was not only the slow time-scale for deployment. Physicists also warned that the United States was moving toward a set of weapons and policies that would rely on relatively inflexible computers. They were particularly concerned with the development of Multiple Independent Reentry Vehicles (MIRVs), which would enable a single missile to send several warheads to geographically dispersed targets. The United States began developing MIRVs in the early 1960s because multiple warheads would deliver destructive power at lower cost, ensure that the offense would penetrate defenses, and enhance a first strike capability.[20]

However, by increasing fears of a first strike, MIRV might also give the Soviets a reason to strike first. Several physicists in the arms control community warned of these instabilities in the mid-1960s, and grew increasingly vocal as the United States began to test MIRV in 1968.[21] Once MIRVs were tested, it would be difficult to negotiate any verifiable limits, because multiple warheads would be hidden within a missile, invisible to satellites. Thus, in June 1969, PSAC's Strategic Military Panel recommended negotiating a freeze on all strategic weapons except for hard-point defenses.[22] Once the MIRV genie was out of the bottle, there would be no option of going back, and the instabilities associated with fears of a first strike would become permanent.

The Nixon administration refused this advice, arguing that MIRV was needed to ensure that U.S. offenses could penetrate the Soviets' defenses. By the same token, the administration argued that Minuteman silos needed Safeguard to defend them against a future Soviet MIRV force.[23]

Herbert York pointed to this logic when he argued: "ABM and MIRV require and inspire each other; together they will lessen our national security."[24] York emphasized that there was more at stake than another expensive lap in the arms race. Because MIRV and missile defenses would "require complex decisions to be made in extremely short times," either a computer or the president would need to "be properly preprogrammed" to decide "whether or not doomsday had arrived."[25]

Indeed, some senators suggested that the survival of Minuteman should be ensured not by the expensive Safeguard system, but by a simple policy: launch the missiles automatically upon warning of an attack. The Nixon administration objected that Safeguard was needed because launch on warning "would not provide the flexibility" needed in the event of a nuclear attack.[26] York went further, arguing that MIRV and missile defenses would encourage "the steady transfer of life-and-death authority from the high levels to low levels, and from human beings to machines."[27]

1.2 Black-Boxing Software

Nixon sought flexibility not just in negotiations with the U.S. Congress and the Soviets, but also in strategic operations. Although the strategic doctrine of flexible response would not be announced until 1974, the administration's interest in flexible nuclear options is evident in his earliest deliberations about missile defense. In fact, some evidence suggests that the Nixon administration wanted to preserve the option of a first strike.[28] When Kissinger evaluated several potential SALT agreements in May of 1969, he worried that the Soviet second-strike capacity might increase with a ban on either MIRV or ABM. Proponents of an agreement argued that this increase didn't matter, since "we will be deterred from attacking" the Soviets, but Kissinger demurred: "I question whether the strength of an American president's resolve in a crisis will be unaffected by the magnitude of Soviet nuclear retaliatory capability."[29]

Significantly, the Nixon administration's hope for more flexible options rested upon systems analysis—calculations that tended to "black box" defensive systems, treating cost as input, and effectiveness as an output. For example, systems analysts often focused on the cost-exchange ratio, calculated by assuming some stable balance of offense to defense, and dividing the cost of one more offensive missile by the cost of intercepting that missile defensively. If the cost-exchange ratio was greater than one, the offense would need to spend more than the defense to maintain equilibrium, and the defense would win, at least economically. Although physicists conducted similar calculations, the economists and mathematicians that dominated systems analysis tended to ignore the ways that the technological design of the system—its particular arrangement of computers, radars, and interceptors—would shape its development and use. And indeed, some of the administration's most vocal supporters in public debate hearkened more from economics and mathematics than physics.

Perhaps most notably, the economist Paul Nitze and mathematician Albert Wohlstetter challenged Wiesner and other physicists who publicly opposed the Safeguard system. As we have seen, Nitze and Wohlstetter had been concerned about a Soviet surprise attack in the early 1950s. The prospect of MIRV only heightened these anxieties, so in the spring of 1969 Nitze and Wohlstetter joined with Dean Acheson to form the "Committee to Maintain a Prudent Defense Policy." They hired three graduate students—Richard Perle, Paul Wolfowitz, and Peter Wilson—to do what Nitze later called the "necessary nitty-gritty of drafting papers to combat the errant nonsense, the inaccuracies and logical tripe being perpetrated by the other side."[30]

While Nitze's and Wohlstetter's used a rigid mathematical analysis, the group largely neglected the internal workings of offensive and defensive systems. For example, they considered only the statistical reliability of individual components, not the less quantifiable risks of "catastrophic" failure. Nor did they analyze the relationship between system design and strategic purpose, for example, which radars, computers, and missiles would be most suitable for hard point defense.

These assumptions led Wohlstetter and his cohort to different conclusions than those of Panofsky and his colleagues. For example, Wohlstetter's calculations showed that "it would cost the Russians more than twice as much to add offense as it would cost us to add an offsetting" defense of missile silos.[31] By contrast, Panofsky argued that the defense would cost more than the silos it defended, as well as "the cost of the enemy missiles which could be intercepted with *confidence*."[32]

This final, emphasized word was crucial. The mathematicians and economists assumed that both sides would have perfect information about the inner workings of offensive and defensive systems, the physicists anticipated uncertainty. Thus Wohlstetter and physical chemist George Rathjens—the former deputy director of ARPA, who had become a professor at MIT—reached different conclusions about the likely effectiveness of a Soviet first strike on Minuteman missiles. Wohlstetter "assumed perfect information would be available to the Soviets" about all of the ways that their missiles would fail, and would be able to retarget appropriately. By contrast, Rathjens assumed that the Soviets, much like the Americans, "would not be able to obtain and use information about such failures in a *timely* fashion but would rather be able to compensate for them at best only on a statistical basis."[33]

Wohlstetter would not accept this answer. In a letter to the *New York Times*, he objected to the notion that "it is safe to wait years for a better ABM," and accused the physicists of being "casual in their calculation."[34] The physicists defended their work, but Wohlstetter would not let the issue die.[35] As the president of the Operations Research Society of America (ORSA), he initiated a study of unprofessional conduct in the ABM debate. The study ultimately censured Wiesner and his cohort for failing to separate the roles of "analyst and advocate."[36] By contrast, it praised Wohlstetter's testimony as "a model for the professional and constructive conduct" of a technical debate.[37]

The study was controversial, even within ORSA. Alain Enthoven dubbed the three MIT physicists "arrogant," and praised the study. But five prominent members of the association, including three former presidents, refused to be associated with the study. For their part, Wiesner and his fellow physicists objected to the premise of the study, for they had never claimed to be operations researchers: "Even the Spanish Inquisition reserved the charge of heresy for those who had been baptized."[38] In a lengthy rebuttal, they noted that the few mistakes in their calculations were "not serious ones," and did not undermine their conclusions, but that the Nixon administration, Wohlstetter, and his committee avoided the more fundamental issues in the debate, "preferring to dwell on minutiae."[39]

Indeed, strategic analyses encouraged a notion of "superiority" that many of the physicists regarded as meaningless. They tended to agree with what McGeorge Bundy, national security adviser to both Kennedy and Johnson, would eventually term "existential deterrence."[40]

Think-tank analysts can set levels of "acceptable" damage well up in the tens of millions of lives. . . . They are in an unreal world. In the real world of real political leaders—whether the United States or the Soviet Union—a decision that would bring even one hydrogen bomb on one city of one's own country would be recognized in advance as a catastrophic blunder; ten bombs on ten cities would be a disaster beyond history; and a hundred bombs on a hundred cities are unthinkable.[41]

Bundy concluded: "In sane politics . . . there is no level of superiority which will make a strategic first strike between the two great states anything but an act of utter folly."[42]

But as Kissinger focused on more narrow mathematical analyses, he grew very concerned about the prospect—however remote—that the Soviets might successfully launch a first strike. He and others in the Nixon administration largely neglected arguments based upon the design of strategic weapons, such as their tremendous complexity and propensity for catastrophic failure.

1.3 Negotiating Limits to Defense

Though arguments about defensive technology did not impress Kissinger, the public debate worried him. Gerard Smith, the head of the U.S. delegation to SALT, recalled: "Nothing concentrated the minds of American leaders on the advantages of SALT as much as the clear and present danger of one-sided arms control in the form of congressional cuts in U.S. defense budgets."[43]

Preparing for negotiations, Kissinger circled the wagons. Nitze's support for Safeguard won him a place as the representative of Defense Secretary Laird on the SALT delegation, while Smith chose other delegates.[44] Kissinger never trusted the SALT delegation or national security bureaucracy, so he set up his own "verification panel" with representatives from various government agencies to discuss options for SALT.[45]

But as Kissinger sought tight control, he began to confront the very *inflexibility* of a missile defense system. By January 1970, the U.S. Congress had committed to deploying Safeguard around two Minuteman silos, and began debating the administration's proposed expansion to five more sites, including the Capitol. In February, the Nixon administration requested support for preliminary work on yet another six sites—one around Minuteman missiles, and five near cities—to prepare for a full, twelve-site system.[46]

Heedless of these commitments, Kissinger authorized the SALT delegation to propose a very different deal to the Soviets: each side would deploy defenses only around their respective capitols, or National Command Authorities (NCA). Kissinger and Nixon believed the Soviet Union wanted a "thin" defensive system, and likely hoped the Soviets would quickly reject the NCA proposal. They were unpleasantly surprised when the Soviet Union expressed interest. If the Soviet position were to leak to Congress, it would undermine support for the administration's proposed twelve-site system.[47]

The administration succeeded in preventing a leak during the congressional debate that year, but it didn't matter. In September 1970, the U.S. Congress authorized defensive sites only around Minuteman bases, not around cities or the District of Columbia.[48] Two months later, the Soviet interest in NCA defenses leaked to the American press.[49] Likely hoping that the prospect of an arms control agreement would sway the Congress, in

January 1971, Nixon renewed his request for a defense around Washington D.C. But without clear congressional support for a defense of Washington D.C., the SALT delegation struggled to retain consistency with their initial proposal to allow only NCA defenses. Kissinger's staff noted: "we are courting a strategic and political disaster."[50]

Thus, the technical features of missile defense limited its political flexibility. The great complexity of defense slowed the deployment, while its geographic dispersal around the nation rendered it vulnerable to local and congressional politics. Political controversy in turn led to continually shifting and ambiguous missions for the system, leaving project managers unable to deliver the requisite technology in a timely manner. In this context, Bell Labs started looking for a way out. A Bell Labs report on programming a Safeguard computer provides some sense of the frustration: ". . . any changes in [the defensive] system objectives and requirements . . . set back the programming. Unfortunately, change and addition are a very real part of ballistic missile defense . . . system design. The result is havoc."[51]

The technical features of strategic weapons also shaped their fate in negotiations with the Soviets. As the public debate continued, the Nixon administration began to see limits on defenses as a good option. Since defensive radars and interceptors could be viewed from space, limits could be verified without on-site inspections. The U.S. Congress also pressed the administration to limit offenses, and especially MIRV. But MIRV was less susceptible to adversarial politics than defenses. Relatively few citizens cared about a few extra warheads in distant missile silos, and it was not until the spring of 1970, that the Senate overwhelmingly passed a resolution urging Nixon to propose a bilateral suspension on deployment of all strategic weapons systems. Nixon called the resolution "irrelevant."[52] In June 1970, the Air Force deployed the first Minuteman III missiles with MIRV, and the very same month, the SALT delegates gave up on negotiating limits to MIRV.[53] The genie was out of the bottle.

The U.S. delegates continued to advocate a comprehensive treaty that would include both offensive and defensive systems. But Kissinger was frustrated with the negotiating impasse and anxious to claim credit for the White House. So in January 1971 he went around his own delegates, initiating "back-channel" conversations with Soviet ambassador Anatoly Dobrynin. Kissinger proposed that they aim for an informal agreement to freeze offensive weapons in exchange for a formal agreement limiting defenses. After several months of discussion, on May 20, 1971, Nixon announced an incipient agreement to negotiate a limit on defenses first, with limits on offensive weapons to follow.[54]

The U.S. delegates were furious that Kissinger had gone around them. Nitze recalled that Gerard Smith talked of resigning in protest. Delegates were also concerned about the damage done by Kissinger's offer. With the Soviets building up offensive missiles, the United States would most benefit from an agreement to limit offenses—and now they were forced to focus on defenses. Nonetheless, they got down to writing the best treaty possible.[55]

Technological features of defense continued to pose challenges for any political agreement. Existing commitments were one source of trouble. The Soviet Union had deployed a rudimentary defensive system around Moscow. By contrast, the U.S. Congress would not tolerate defenses around Washington D.C., but had committed to deploying defenses around Minuteman silos. Kissinger suggested a trade-off that would allow both sides to keep systems designed for different purposes, but to no avail. Eventually, a Soviet delegate and radar expert, Alexander Shchukin worked with Smith and Nitze to break the impasse. They agreed to limit technological components—missile launchers, interceptors, and radars—and their geographic distribution, since this in turn would limit the purpose of the defense.[56]

Negotiators also grappled with an uncertain future. Scientists and engineers were growing increasingly hopeful about the future of laser weapons—technologies that they had long hoped would kill missiles at the speed of light.[57] Efforts to limit lasers and other future technologies raised complex questions about the distinction between research, development, and deployment. When Kissinger's verification panel tackled these issues in early August 1971, the discussion was "fragmentary and not very orderly."[58] The chairman of the Joint Chiefs of Staff, Admiral Thomas H. Moorer, did not want any restrictions, arguing that limiting deployment would also limit important research and development. Deputy Secretary of Defense David Packard felt that they could foreclose deployment while allowing "the necessary research and development."[59]

Nixon favored Packard's proposal.[60] But the notion of "development" remained ambiguous, and Shchukin asked the United States to clarify. Brown replied that while research would include "conceptual design and laboratory testing," development would include "the construction and testing of one or more prototypes of the weapon system or its major components."[61]

Though the SALT delegates nearly finalized the treaty in the spring of 1972, Nixon and Kissinger tinkered with the wording until the end. The White House wanted to claim credit for the agreement, and excluded its own delegation from most of the summit in Moscow, finally staging a high-drama signing on May 25, 1972.[62] The United States Congress overwhelmingly ratified the treaty in August 1972.[63]

The final treaty limited each country to defensive deployments at no more than two sites—one around their respective national capitals, and another around ICBM silos—and imposed limits on the numbers of radars, missiles, and launchers each could deploy in each area. The United States and the Soviet Union also agreed not to "develop, test, or deploy" defensive "systems or components which are sea-based, air-based, space-based, or mobile land-based." It also forbade deploying more advanced missile interceptors and launchers. The treaty did not explicitly forbid research or testing of "ABM systems based on other physical principles" but agreed that if they were created, limitations would be subject to review by a standing consultative committee.[64] As we will see, a future administration would challenge even this constraint.

1.4   Software and the Evolution of Defense

Although the ABM treaty said nothing about software or computers, the public debate and growing concerns about slow software development had a long-term impact on the Army's missile defense effort. Software became increasingly difficult to manage in the late 1960s and early 1970s, as the changing mission of missile defense forced almost continual reorganizations. Anticipating the deployment of Sentinel, Foster felt the Army needed a "head graft," so near the end of 1967, he abruptly transferred most of ARPA's Defender program to the Army's new "Advanced Ballistic Missile Defense Agency."[65] A few years later, the Army created organizations to oversee training, logistics, and other aspects of Nixon's proposed 12-site Safeguard deployment. But, in 1972, the ABM treaty limited Safeguard to a single site in North Dakota, so the Army was forced to consolidate management into a much smaller "Ballistic-Missile Defense Organization."[66]

By 1973, the Army's effort consisted of three programs—Advanced Technology, Site Defense, and Safeguard—which spanned the spectrum from research to development and deployment. The Advanced Technologies Program researched novel technologies, such as lasers, sensors, and computer architectures. Site Defense developed technologies that could be rapidly used in response to an evolving Soviet threat. And the Safeguard program was deploying defensive battalions.[67]

Although all three programs worked with software, the Site Defense program put it front and center. The Site Defense program emerged in 1970 to counter physicists' arguments that the complex software would render Safeguard obsolete by the time it was deployed.[68] Foster explained that Site Defense would address the challenges of "maintaining a schedule" and obtaining "cost effectiveness" in the unprecedentedly complex software required to integrate a missile defense system.[69]

Software also became the central means of testing whether Safeguard "worked." In its 1973 review of Safeguard, the General Accounting Office noted that a "complete determination of system effectiveness cannot be made by testing alone," because of "testing limitations and constraints." Hence Safeguard was assessed "through a combination of computerized simulations, engineering analyses, and actual tests."[70] The Army intended to continue using simulations to practice its operations after the system was deployed, but by 1975, the Army was also quietly discussing plans to scale back operations, recognizing that the deployed system would soon be obsolete.

On October 1, 1975, the Army declared Safeguard operational. The next day, having learned of the Army's plans, the House of Representatives voted to "close down Safeguard, tear it out and ship it off."[71] Ultimately, Congress passed a more moderate measure: nuclear warheads and missiles would be removed, and the missile site radar would be shut down to save operating costs, but the long-range Perimeter Acquisition Radar (PAR) would be left intact and integrated into the U.S. early warning system.[72]

The closure of Safeguard was strikingly anticlimactic, following the "military maxim that outmoded weapons should not be killed but should just fade away."[73] Residents of Langdon, North Dakota, whose economy had boomed to support the Safeguard deployment, were among the few mourners. When the construction of Safeguard began, schools, stores, and housing grew to support construction workers, technicians, and engineers from across the country; at one point the schools included children from 48 states. The town's part-time mayor explained: "We knew about the bargaining-chip theory a couple of years ago. But it's been here so long now, that we assumed it would keep going."[74]

Without Safeguard, the Army's missile defense program consisted only of the Advanced Technologies Program and Site Defense. The latter program became the Systems Technology Program (STP) and increasingly focused on software. One Army official explained that STP would address "one of the weaknesses" in early efforts at missile defense: "We built interceptors, we built radars, but we really didn't pay much attention to how you tied them together."[75] He noted that in "most weapons systems the software is relatively routine, contrasted with ballistic missile defense where it tends to be the essence of the system."[76] Since the Army could not test a fully integrated missile defense system, it used simulations to check "the complex interactions of subsystems, components, and environments in candidate BMD systems of the future."[77] As we will see in the next chapter, the Site Defense and Systems Technology Programs contributed to the rise of "software engineering" in the 1970s.

2   Public Debate and the Shifting Place of Physics

2.1   New Modes of Political Engagement

For many of the president's science advisors, the missile defense debate was part of a broader shift in modes of political engagement. Many grew increasingly frustrated that they were not consulted on important decisions. Furthermore, even when the Nixon administration solicited their advice, officials in the administration misrepresented the advisors'conclusions.[78]

For example, when senators asked Packard whether he had consulted with experts outside of the Defense Department, he mentioned Panofsky's name—despite the fact that Panofsky disagreed with the administration's decision. However, a staffer noticed that Panofsky happened to be in the audience. Panofsky recalled: "Gore fingered me and said, can you testify tomorrow? And so I got gored."[79] Later that week, Panofsky testified that he "did not participate in any advisory capacity" in reviewing the U.S. missile defense program, though he acknowledged "an informal discussion with Mr. David Packard."[80] Gore asked: "Was there an extended conversation over a period of time?" Panofsky replied: "About half an hour. . . We happened to accidentally meet at the airport."[81]

In 1970, John Foster went further. In congressional testimony, he described an ad hoc study for which he "deliberately chose scientists who opposed the deployment of the Safeguard as well as those who favored it."[82] Foster reported that he had instructed the panel "to put politics aside and just ask the question: Will this deployment, with these components, do the job that the Department of Defense is trying to do?"[83] According to Foster, the panel answered yes.

But two panel members, Sidney Drell and Marvin Goldberger, immediately wrote letters to object, and were summoned to testify. Goldberger explained: "Although we had been told when invited to join the panel that we would be presented with the Department of Defense recommendation . . . and that any objections we raised would be passed on to the administration, this did not in fact happen."[84] Instead, he continued, the panel was asked to consider a variety of options: "Since there was no specific deployment proposal presented to the panel . . . we could scarcely have concluded that '. . . this equipment will do the job that the Department of Defense wants to do.'" [85]

Drell reiterated these points, calling Foster's statement "incorrect," but concluded: "I want it to be absolutely clear that I do not intend in any way to impugn Dr. Foster's integrity by these remarks." This irked Fulbright: "What do you intend? You just said he is a liar." Drell insisted that Foster must have misremembered because of his "very many important responsibilities."[86] Fulbright pressed the point, to no avail. He finally stormed out of the room in frustration.[87]

Similar controversy emerged when Packard attempted to discount York's testimony. Packard suggested that York once opposed the Polaris subma-rine- launched ballistic missile system, which by 1969 was accepted as the most invulnerable part of the strategic triad. York soon telegrammed a correction: ". . . I always during my tenure with the Defense Department recommended positively to the Secretary of Defense concerning the devel-opment and deployment of the Polaris system." [88]

These exchanges cast doubt on the integrity of the Nixon administra-tion, even as they contributed to rifts among physicists. Privately, York blamed the press for introducing "personal elements" into the public dis-agreement.[89] He offered a sympathetic explanation:

I understand from those who know [Packard] that he really is a good and intelligent man. I think the trouble is that he was forced by political events to take a stand on the ABM which simply was not and could not be based on sufficient analysis on his part. He had not even been Deputy Secretary of Defense for two months, and has been faced with hundreds of other new and difficult problems at the same time. As a result, he is being harassed by events and is responding in a flustered manner.[90]

York, Panofsky, and other elite physicists walked a tightrope as they sought to influence public debate without losing inside influence. In Sep-tember 1969, Drell asked Wiesner for help in publicly opposing what he called "Johnny Foster's outrageous testimony."[91] He explained that despite a public emphasis on deterrence, the administration was "working toward a damage limiting strategic capability"—something that would leave the Soviets fearful of a first strike.[92] Drell noted several issues that would be best fought through "internal discussions in Washington," but in other areas believed that "the credibility of the Foster-Laird position and statements should be attacked in public."[93] He believed "it would be useful to make clear both to the Congress and interested citizens how extreme (paranoid!) a view of the strategic arms race these gentlemen hold."[94] He continued: "I am turning to you because you have the contacts, and moreover I have to tread carefully here because of my present direct activities in this very area through SMP [Strategic Military Panel]."[95]

Other physicists were less circumspect. For example, Garwin publicly opposed the supersonic transport (SST), which had begun in 1963 as an effort to build the world's best commercially successful supersonic airplane. PSAC warned of problems—noise pollution and environmental damage due to sonic booms, the inadequacy of existing runways, the inefficiency and associated questionable commercial viability of the plane—but the project continued. When Nixon inherited the SST in 1969, he asked the Department of Transportation to coordinate a multi-agency review of the program. DuBridge obtained Nixon's approval to form a separate "ad hoc SST Review Committee," and put Garwin in charge.[96]

Without publishing any of the studies, Nixon announced in September 1969 that the "SST is going to be built," and that he would request congressional funding for two prototypes.[97] But he encountered immediate opposition from congressional critics, who sought access to the White House studies. They obtained the Department of Transportation report, and discovered that it was highly critical of the SST. They also sought Garwin's study, which had gone beyond critique, recommending that the SST be canceled. But Nixon prevented release of Garwin's report, claiming executive privilege.[98]

Hence, in the summer of 1970 Garwin appeared before the Senate subcommittee on transportation to testify "no longer as advisor, but as citizen."[99] He brought a briefcase bulging with charts and other analyses that—as he repeatedly emphasized—he had produced himself. He testified that the SST would produce the same amount of noise as 100 jets taking off simultaneously, and that the program should have been canceled more than a year earlier.[100] In December 1970, the U.S. Congress refused to fund the SST.[101]

The ABM and SST debates forced physicists to think hard about just how and when to go public. As Nixon's full-time science adviser, DuBridge always supported the president's decisions publicly, even when it cut against PSAC advice—a choice that alienated members of PSAC. Wiesner felt that full-time advisors should not oppose the president: "If you become so disaffected with its programs that you want to fight, the proper thing to do is quit."[102] However, he also felt that part-time advisors "have much less obligation."[103] He acknowledged that "it's still regarded as bad taste to engage in public debate" and that some "younger people," such as Garwin, had been criticized, but felt "that the nation has paid a much higher price for its secrecy than it would have paid through a policy of complete openness."[104]

2.2   "The End of an Era"

Although Wiesner and other elites advocated greater openness, they also worried about a loss of influence—and rightly so. After a term marked by major controversy, presidential advisor Lee DuBridge tendered his resignation in August 1970. As *Science* noted, DuBridge had alienated "an influential segment of PSAC which felt that he had violated PSAC etiquette by voluntarily speaking out in support of the administration's antiballistic missile system."[105]

Science also noted that Nixon's chosen successor to DuBridge, Edward David, represented a "break from the past" in two significant ways.[106] First, as an engineer at Bell Labs, David was "the first to come to the science adviser's post from industry rather than the university."[107] And second, he had not been "initiated into the public service in the mobilization of scientists in World War II."[108]

Indeed, at 45 years old, David was 23 years junior to DuBridge. He had never served as a member of PSAC, though he was familiar with its members—he had earned his doctorate in electrical engineering at MIT, and remained an associate of Wiesner and others at MIT. As we have seen, David was sufficiently established in computing to be a participant in the first software engineering conference, but this was a new and unknown field. As *Science* noted, "David is well known in his own field, but his name is hardly a household word in the scientific community."[109] David's appointment signaled "the end of an era in science affairs in this country."[110]

But far more radical changes were on the way. A year had hardly passed when Nixon revealed that he would eliminate PSAC and the Office of Science and Technology as part of a broader restructuring in the federal government. On January 2, 1973, David resigned in anticipation of the changes, and the rest of the PSAC members tendered their resignations soon after.[111]

By early February, a new plan was emerging: the director of the National Science Foundation, Guy Stever, would help lead scientific advising, reporting to Nixon's "economic czar," George Schultz, through a lawyer. *Science* noted that science policy would "now be made not by top-ranking scientists but by young Republican lawyers."[112]

Physicists also grew concerned about a loss of government support for their research. In August 1969, Senator Mike Mansfield (D-MT) proposed an amendment to the 1970 Defense Procurement Authorization Act: projects could be funded by the Defense Department only if they had a "direct and apparent relationship to a military function or operation."[113] The Senate had discovered that the Defense Department was representing basic research as projects with a defense application, and demanded more accountability. But the amendment produced immediate disarray, as the distinction between "basic" and "applied" research was fuzzy. To clarify matters, the 1971 bill was revised to give the secretary of defense discretion over what counted as "defense-related" research. Mansfield felt that this only worsened matters.[114]

Perhaps nobody was more aggrieved by the Mansfield amendment than physicists, who had relied heavily upon Defense Department funding. Although some projects were transferred from the Defense Department to the National Science Foundation and other civilian agencies, funding still fell short. Moreover, the amendment encouraged managers in the Defense Department to fund projects with an immediate application rather than more basic research. Eberhardt Rechtin, who was director of ARPA when the amendment passed, called it "a disaster" because it made the bureaucracy "*extremely* conservative."[115] He explained that program managers "wouldn't put money out to a university for anything, unless it was an obvious need to protect a tank," but with student protests on the rise, "universities didn't dare say that they were working on combat arms."[116] Virtually all the scientists who had helped direct research policy, including the president's science advisors, agreed that the amendment was "*very* bad" for government funded research.[117]

A few scientists embraced the amendment as a way of redirecting research priorities. As Kelly Moore has discussed, some scientists grew increasingly critical of military researchers, and sought professional reform.[118] For example, in 1968, UC Berkeley physics professor Charles Schwartz proposed to amend the American Physical Society (APS) constitution so that the association could pass resolutions with social implications. The general membership defeated the proposal by a two to one vote. But Schwartz and a few other supporters formed "Science for the People," a radical group that grew increasingly disruptive, suggesting that physicists who had consulted for or worked on weapons-related research had committed war crimes. Activists in the physics community cheered the Mansfield amendment. Cornell physics professor Jay Orear argued that objections to Mansfield came from "the establishment-controlled Council" of the APS, not "most of our younger physicists."[119] Orear noted "a strong protest movement within the physics community whose goals and tactics have been rapidly evolving over the last few years," and predicted "that this movement is here to stay."[120]

Indeed, the antiballistic missile debate spurred many scientists to institutionalize new modes of political engagement in the 1970s. As we have seen, elite physicist-advisors and somewhat younger activists came together in their opposition to missile defense, organizing "teach-ins" at universities across the nation on March 4, 1969. The faculty associated with MIT's teach-in established the Union of Concerned Scientists, a group that began to oppose nuclear power as well as nuclear weapons.[121] In 1972, the younger physicists helped establish the APS "Forum on Physics and Society" to study "the interrelation of physics, physicists and society."[122] Two years later the APS established a more elite group, the Panel on Public Affairs (POPA).[123] As an advisory group elected by the APS Council, POPA became the society's means of assessing issues in the public arena. As we will see, these groups would provide new ways of opposing presidential policy when public debate over missile defense debate rose again in the 1980s.

3 Physics and Computing in the New Political Establishment

In sum, the missile defense debate was part of a broad set of transformations in the relationship between and among policymaking, physics, and computing in the late 1960s. The ABM treaty was neither the natural result of obvious technical realities, nor was it a purely social construct. The political debate contributed to a constantly changing mission for missile defense, exacerbating the challenges of developing complex software. Equally, the technological features of defense—complex software, geographic distribution, and visibility—shaped the negotiation of the ABM treaty.

Furthermore, the debate carried implications for far more than the ABM treaty. Anticipating the treaty in the summer of 1971, physicists reflected on whether they had won or lost the debate, but could find no simple answer. The Army had begun defensive deployments that would soon be obsolete. Yet Kistiakowsky felt that "in a more fundamental sense we won, because we generated a completely new phenomenon. . . . The proposals of . . . 'hot-rod military' types, are not sacrosanct anymore." Wiesner went further: "I think in a real sense the nation won. Congress looks at everything seriously now."[124] Indeed, as we will see in the following chapter, congressional oversight of Defense Department projects grew throughout the 1970s.

But whatever their successes, physicists felt a loss of influence in the years following the Safeguard debate. Not only did Nixon abolish PSAC, but the Mansfield amendment contributed to a shrinking academic discipline. As measured by the production of PhDs, physics slowed dramatically during the 1970s. What David Kaiser has termed the physics "bubble" had burst.[125]

By contrast, academic computer science grew steadily in the 1970s. It was buoyed by a reorganization of the National Science Foundation. In the 1960s, computing was seen primarily as service to other disciplines, and thus about 30–40 percent of the NSF computing budget was devoted to developing computer applications for other fields. But in the mid-1970s, the NSF transferred work on computer applications to the disciplines they aimed to serve, while holding the "computer science" budget constant—effectively expanding support for computing.[126]

Meanwhile, concerns about software for missile defense contributed to a growing investment in software research and development. More broadly, growing congressional oversight spurred the Defense Department to search for a solution to expensive and glitch-ridden software. As the next chapter shows, that solution took the form of "software engineering."

7   The Political Economy of Software Engineering

"Is software engineering a fad?" . . . the field appears to have largely sprung up in the last two years. . . . Since then, any activity or organization with the words "software engineering" (or something vaguely related to that) has had a virtual guarantee of success.

—Tony Wasserman, *Software Engineering Notes*, April 1977

In 1977, the chairman of a young Special Interest Group on Software Engineering, Tony Wasserman, confronted the sudden rise of his field. He felt that software engineering "has attracted a number of people who think (or hope) that it will provide a panacea for all of their problems in software production . . . this first wave will certainly drift away." Nonetheless, he believed that "the discipline seems like it is here to stay" since it promised to meet crucial needs, ranging from certifying "the correctness of software products," to "teaching software techniques to students and professionals."[1]

In many ways, Wasserman's remarks proved prescient. The journals, special interest groups, and conferences that began in 1975 continue today, nearly forty years later. Yet the persistence of software engineering is puzzling. The first NATO-sponsored Software Engineering Conference in 1968 was filled with optimism, but a second such conference in 1969 was bitterly divisive. The disastrous second conference is commonly understood as a sign of lasting divisions in the field.[2] NATO never sponsored a third conference, and many of the goals outlined by Wasserman—such as certifiably correct software, or educational requirements for professional software engineers—remain elusive today.

How, then, did software engineering get established? What drove the formation of the field? And how did practitioners of software engineering come to understand the challenges and risks of complex software?

This chapter shows how the meaning, practice, and status of software engineering changed through the 1970s.[3] In 1969 "software engineering" did not connote any formalized practices or knowledge. But by the early

1980s, software engineering was institutionalized throughout industry, government, and academia. International journals, professional groups and conferences had formalized methods, knowledge, and practices. Software engineers had established a disciplinary repertoire.

To understand these changes, I argue that we must consider the shifting political and economic contexts of computing. As Martin Campbell-Kelly has discussed, analysts in the computing industry came to believe that software costs were on the rise in the late 1960s and early 1970s.[4] These concerns were also worrisome for a major source of funding in the computing industry—the U.S. Defense Department—which felt growing congressional scrutiny and shrinking budgets. Widely publicized problems in command and control systems underscored weaknesses associated with fragmentation—the military services used dozens of incompatible computers and languages—while budget cuts prohibited urgently needed upgrades. Worst of all, Pentagon managers confronted growing demands for responsiveness and interoperability. Throughout the 1970s, the Defense Department sponsored software engineering research in an effort to cut costs, and thus fostered a specific vision of software engineering—as a means of quantifying, predicting, and managing complex software. The sheer scale of Pentagon-sponsored software engineering gave this vision influence well beyond military projects.

Many software researchers were wary of the managerial interests that drove funding for the new field. They feared that the automation of programming would replace "intellectual discipline by management discipline," and noted the persistence of errors in nuclear command and control systems.[5] Furthermore, the shortcomings of management grew increasingly evident with the rise of the microprocessor industry. Not only were the new devices programmed by amateurs, but their users were not trained and managed like business and military computer users. As computer software was embedded in complex systems, ranging from pacemakers to public transportation systems, many software researchers began to express concern about the risks of large, complex technological systems. Ironically, efforts to establish software engineering helped produce a new research agenda—one that was primarily concerned with the fallibility of complex systems.

1   Making Software Engineering Persuasive

1.1   A False Start
Divisions in software engineering emerged not long after the first software engineering conference, in distinctive notions of "structured programming." The 1968 conference helped inspire the Dutch computer

scientist Edsger Dijkstra to write *Notes on Structured Programming*.[6] The *Notes* extended and formalized some ideas that Dijkstra had expressed in a March 1968 letter to the editor of *Communications of the ACM*, published under the title "Go-To Statement Considered Harmful." There, Dijkstra argued that a common instruction—telling the computer to "go to" a different line in the program—had "disastrous effects." He pointed out that a computer program was both code viewed by programmers in *space*, and a set of processes run by a machine in *time*. Since human "powers to visualize processes evolving in time are relatively poorly developed" programming would be improved by making the program's temporal process visible in the code. By contrast, the "unbridled use of the go to statement" made it "terribly hard to find a meaningful set of coordinates in which to describe the process. . . ."[7]

Dijkstra's letter began a long tradition of articles warning of methods that were "considered harmful."[8] Many in the computing community feared that Dijkstra intended to structure not just computing code, but also to rigidly structure the work of programmers.[9] In fact, Dijkstra was on the side of programmers, not managers. In *Notes on Structured Programming*, Dijkstra provided a much more general and formal analysis, showing how programmers might structure their code like mathematical theorems, which could be proven correct.[10] Dijkstra's *Notes* were circulated widely, and "structured programming" soon became a widely touted approach to software development.

Industrial developers advocated a different, managerial conception of structured programming. For example, at IBM, the distinguished mathematician and computer scientist Harlan Mills argued that a unified, "top-down" approach to program design would improve efficiency. Pointing to a software project that was then being managed in a traditional "army of ants" fashion, Mills proposed a race—he would "do in six man months what was regarded as being essentially a 30 man year project." Initially dubbed a "super-programmer," Mills became a "chief programmer"—someone who required a small staff of clerical and technical assistants. Using Mills's approach, the project required closer to six man years.[11]

Others at IBM likened Mills's approach to "a surgical team in which chief programmers are analogous to chief surgeons, and the chief programmer is supported by a team of specialists. . . ."[12] Such top-down structured programming would allow programming to proceed more "effectively" and "economically."[13]

Though the notion of a surgical team was intended to be flattering, many researchers resented the emphasis on managerial control. When IBM manager J. D. Aron presented these results at the second software engineering conference in 1969, he highlighted the value of "a management and control organization that makes the best out of the resources that are normally available." John Buxton, a professor at the University of Warwick, objected that software was not "completely and totally a management problem," and implied the need for better programming languages and techniques. Aron agreed, but insisted that "management decisions are required to take advantage of any technology that is available."[14]

Alan Perlis of Carnegie Tech countered that the managers were out of their league, citing the Job Control Language (JCL) developed for OS/360:

I have yet to hear a good word about JCL. What kind of managerial structure is it that doesn't filter a new proposal against existing know-how so that, for example, JCL would have the right properties of a language? It is failures like that which are technical failures on the part of managers, and which are largely responsible for major difficulties in the system. I think that, by and large, the managers wouldn't know a good technique if it hit them in the face.[15]

Perlis and like-minded researchers were much more interested in Dijkstra's structured programming, as he aimed to show how programs could be formally proven correct. Dijkstra acknowledged that his experiments were "rather small," but insisted that his "real concern" was large and complex programs.[16]

Edward E. David—the Bell Labs engineer who would soon become Nixon's science advisor—noted that such small-scale experiments were not necessarily relevant to the real world. He suggested that software engineers should instead study "actual ongoing production efforts" to gather more "evidence and hard data about real cases."[17] Dijkstra objected to any such distinction "between practical and theoretical people," calling it "obsolete, worn out, inadequate and fruitless," continuing: "I absolutely refuse to regard myself as either impractical or not theoretical."[18] He accused managers of resisting technological improvements. His audience applauded.

The editors of the conference proceedings concluded that divisions between management and technically oriented people were "a reflection of the situation in the real-world," and that "the realization of the significance and extent" of these divisions was "the most important outcome of the Rome conference."[19] However important this realization, conferees experienced "an enormous sense of disillusionment," a "bad-tempered" conference, a "disaster."[20] For at least another five years, very few books or reports would feature software engineering (see figure 7.1). To understand how interest in software engineering resumed, we must consider how the economic and political environment changed in the years following the conference.

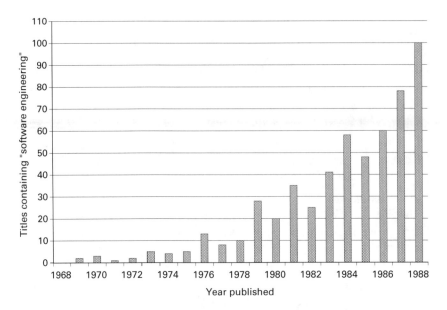

Figure 7.1

Number of books or reports with the word "software engineering" in the title, by year of publication. Data were gathered from the WorldCat.org database.

1.2 Accounting for Software: Rising Costs, Shrinking Budgets

In the 1970s, several changes in the computing industry contributed to growing alarm about the rising costs of software. One dramatic change came from industry giant IBM, which announced in 1969 that it would no longer "bundle" software services with its hardware—henceforth, it would charge for each computer program or package. For years, industry analysts had viewed software as a dominating cost of products like OS/360, and believed IBM should sell its hardware at a much lower cost than its bundled systems. Instead, IBM lowered its prices by only 2–3 percent. Thus, most analysts expected the effective cost of software to rise sharply.[21]

Additionally, time-shared computers that could be accessed by multiple users in real time were starting to look expensive. The cost of time-sharing software—programs to run terminal displays, keyboards, and printers—seemed to be growing. Werner Frank, a cofounder of the software services firm Informatics, predicted in 1968 that the ratio of software to hardware expenses might rise to 80 percent of the total cost of computer systems by 1978.[22] Authors plagiarized Frank's plot throughout the 1970s, retracing the shape but shifting the dates ever farther into the future. Predictions of growing software costs, combined with the unbundling decision, nurtured the rapid growth of a "packaged software" industry.[23] Years later, Frank would dub this "Myth No. 1," pointing to difficulties in distinguishing the costs of various stages of programming.[24] (See figure 7.2.)

Costs were also worrisome to managers of government software.[25] In the wake of 1960s controversies about missile defense and military policies, the U.S. Congress increased scrutiny of Defense Department spending, demanding accountability and greater effectiveness, and software became a target of focused critique.[26] A Nixon-appointed blue ribbon panel on defense acquisition reform in the summer of 1970 underscored the need to account for total lifecycle costs—development, production, maintenance, and operations.[27] In the arena of computing, software was the dominant lifecycle cost—programming, debugging, testing, and upgrading—and rising costs were worrisome. The panel also criticized the Defense Department because rather than time-sharing machines, it duplicated computers, costing an additional $500 million per year. Its computers were quickly growing obsolete.[28]

Perhaps nobody struggled more with these challenges than those charged with managing computerized command and control systems.[29] The basic challenges were well recognized by the mid-1960s, when *Armed Forces Management* noted that the large number of agencies involved with command and control, combined with a "lack of clear-cut responsibility" and constantly "changing requirements," had produced "systems that were enormously expensive and relatively ineffective."[30]

A wide range of incompatible computer systems could not easily be integrated into the National Military Command System (NMCS) that McNamara had ordered in the early 1960s. Nor could they evolve with strategic policies. As chapter 3 discussed, the Air Force–commissioned "Winter Study" had underscored the need for an "evolutionary" approach to command and control, and for standard languages, as early as 1959. Yet by 1969, when the NMCS had become the World Wide Military Command and Control System (WWMCCS), it encompassed 131 different computer systems and 23 programming languages for 74 distinct command missions.[31] The lack of standardization not only undermined the performance of WWMCCS as a unified system; it also increased programming and training costs.[32]

In November 1969, the new deputy secretary of defense David Packard continued McNamara's efforts at unifying command and control by announcing that the Defense Department would purchase 34 standard computers through a competitive bidding process, with up to 35 more to follow. Officials outlined a five-year plan to "standardize computer operations," emphasizing "software, programming and support."[33] In principle, the new computers would include time-sharing and cut lifecycle costs, as recommended by Congress.

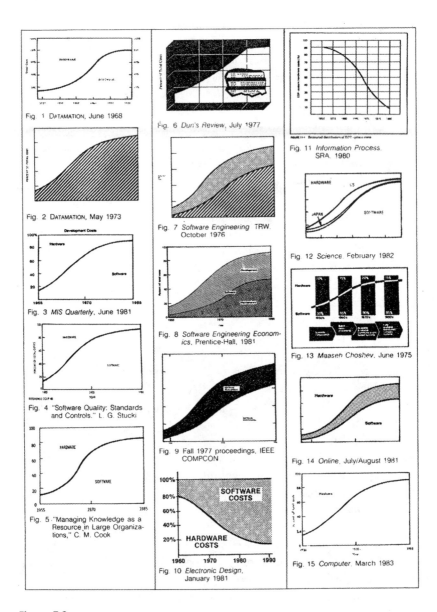

Figure 7.2
Expectations of rising software costs, as illustrated by several very similar charts, and debunked as "Myth No. 1." Reprinted with permission of Werner Frank.

Budget cuts soon undermined efforts at reform. By June 1970, Packard was forced to scale back his proposal to 15 new computers, with the option to buy 20 more.[34] In November 1970, the House of Representatives again delayed funding for the WMMCCS upgrade while the General Accounting Office (GAO) slowly reviewed the system's management. When the GAO finally sent its report to the House Appropriations Committee in early 1971, *Datamation* observed that the upgrade "appeared to be more dead than alive." The GAO accused planners of using "highly questionable" cost estimates and allowing "fragmented responsibility for planning and direction of acquisition." It also criticized the Pentagon's cost estimates for the upgrade—which would include $722 million for software, $206 million for hardware, and $31 million for management—demanding better documentation.[35]

In January 1971, *Government Executive* anticipated that the ongoing program reviews had added a year or more to the WWMCCS upgrade schedule.[36] Worst of all, in October of 1971, the Defense Department issued a contract for 15 computers that were already growing obsolete. The computer hardware was *not* designed for time-sharing—a decision the Air Force had protested internally, to no avail. Plans for a real-time computer network were pushed into the future, as the Defense Communications Agency began work on a "Prototype WWMCCS Intercomputer Network" (mercifully shortened to PWIN).[37]

1.3 Managing Command-Control Software: Fragmentation and Flexibility

At the same time that Defense Department managers faced heightened scrutiny and budget cuts, they were confronted with shifting policies that would require command and control systems to become more "flexible." Yet as three highly visible failures demonstrated, the WWMCCS was anything but flexible and responsive.

On June 8, 1967, during the six-day Arab-Israeli war, a Navy research vessel, the USS *Liberty*, came under friendly fire from the Israeli Navy and Army—killing 34 crew members. U.S. commanders had sent five messages instructing the *Liberty* to move to safer waters, but "communications delays and misrouting errors" prevented the messages from getting through.[38] Just six months later, a Navy spy ship, the USS *Pueblo*, was boarded and captured by North Korean forces as it operated off the coast of Japan. "Grossly excessive" delays in transmitting the *Pueblo's* distress messages prevented U.S. forces from responding. One sailor was killed, the entire crew was long imprisoned, and the valuable ship was permanently lost.[39] Finally, late on

April 14, 1969, a U.S. tracking station in South Korea noted that North Korean aircraft were tracking a Navy reconnaissance plane, the EC-121 as it flew over the Sea of Japan. The station called for military help, but its four messages were delayed by well over an hour. The next set of messages reported the likely shoot down of the plane—and indeed, all 31 Americans on the spy plane were killed as it crashed off the North Korean coast.[40]

The House Armed Services Committee investigated these failures, concluding that "fragmented and overlapping responsibility" for the development of WWMCCS had produced "ineffective management," cost overruns, and an unresponsive command and control system.[41]

Pentagon managers were still wrestling with these problems when they began confronting demands for unprecedented levels of flexibility and responsiveness from WWMCCS. In January 1974, Nixon issued a National Security Decision Memorandum, seeking "a more flexible posture," directing the development of "a wide range of limited nuclear employment options."[42] When Nixon's secretary of defense, James R. Schlesinger, announced the decision, he sparked considerable debate about whether Nixon viewed limited nuclear war as winnable.[43]

Far less attention was devoted to a gap between official policies and technical development. The Nixon administration focused almost entirely on improving physical systems rather than information systems. For example, when Kissinger met with the verification panel to discuss options for controlled nuclear escalation in August 1973, he focused almost entirely on gaining congressional approval for new weapons. Fred Ikle, representing the CIA, pointed to a different problem: command and control vulnerability.[44]

Spending on physical elements of command and control—sensors, radars and computers—grew significantly for each of the services in the 1970s.[45] For example, the Air Force upgraded the Satellite Early Warning System and Submarine Launched Ballistic Missile system, replacing dish radars with phased arrays.[46] By contrast, efforts to unify the services' disparate command and control systems by standardizing software and other aspects of operations suffered. The services had relatively little incentive to invest in joint programs—which could easily be seen as competitive with individual programs. In 1970, David Packard created a new office, the Assistant Secretary of Defense for Telecommunications, with the authority to require more inter-service cooperation.[47] But just two years later, Schlesinger reduced the authority of the position, moving it to the obscure Office of Legislative Affairs.[48] In his pursuit of flexible options, Schlesinger instead focused on developing a more accurate missile for the Air Force, the MX.

The demand for more flexible command and control systems continued under President Jimmy Carter, who replaced the notion of "flexible deterrence" with "countervailing."[49] Carter's administration managed command and control systems more carefully than Nixon's. In 1977, the new secretary of defense, Harold Brown, consolidated two positions—the Director of Telecommunications and the Assistant Secretary of Defense for Intelligence—into one more powerful position, the Assistant Secretary of Defense for Command, Control, Communications, and Intelligence (C3I). To further integrate management, the new secretary also reported directly to William J. Perry, the director of defense research and engineering.[50]

Unfortunately, these improvements could not instantly reverse decades of management problems. By 1977, John H. Bradley, a computer scientist in the DCA, was struggling to link the WWMCCS computers into a prototype network. The computers were not designed for real-time operations, but Bradley's efforts to highlight problems were ignored. He went over the heads of his superiors, and eventually went public. Bradley was fired in June 1977, prompting more publicity and an investigation by the General Accounting Office.

In 1979, the GAO issued a report blasting the WWMCCS management for producing a system that was so "complex and fragmented" that it was unreliable, inefficient, and could not be "responsive to national or local level requirements," let alone grow economically or effectively. Indeed, the decision to purchase unwieldy, obsolete computers encouraged "independent and decentralized software development efforts," as users wrote their own patches to work around the official systems software.[51]

1.4   Software Engineering as Management Science

Thus, managers of Defense Department computing found themselves between a rock and a hard place. While anticipating the rising costs of software, they also confronted shrinking budgets, perpetual scrutiny, and growing demands for flexibility, interoperability, and security. In this environment, defense department managers began to embrace a particular vision of software engineering, a means of quantifying software development and ensuring accountability.

The managerial version of software engineering was well articulated in a 1970 Air Force study of future Command, Control, and Information Processing needs. Commonly known as CCIP-85, the RAND-led study aimed to provide a 15-year road map for future research and development.[52] The head of RAND's Computer Systems Analysis group, Barry Boehm, took on the CCIP-85 study and soon became one of the most influential advocates of software engineering.

In 1973, Boehm summarized one of the study's primary conclusions: software would consume 90 percent of the Air Force's Advanced Data Processing budget by 1985. He explained:

... for almost all applications, software (as opposed to computer hardware, displays, architecture, etc.) was the "tall pole in the tent" —the major source of difficult future problems and operational performance penalties. However, we found it difficult to convince people outside of the software business of this . . . because of the scarcity of solid quantitative data to demonstrate the impact of software . . . [53]

To gain credibility, Boehm sought as much quantitative data as possible. For example, he estimated that the Air Force had spent at least $1 billion on software in 1972, roughly 5 percent of the entire Air Force budget. He also noted congressional critiques of WWMCCS.

Boehm highlighted "problems of management."[54] Because experience was not transferred from one project to the next, "lessons learned as far back as SAGE [10 years earlier] are often ignored in today's software developments. . . ."[55] Boehm argued: "Until we establish a firm data base, the phrase 'software engineering' will be largely a contradiction in terms." He quoted Lord Kelvin: "When you can measure what you are speaking about, and express it in numbers, you know something about it."[56]

In 1973, Boehm took this vision of software engineering to TRW, a major defense and aerospace contractor, where he became the director for software research and technology. Boehm recalled that TRW's contract to develop software for Site Defense—the Army's prototype program for defenses that could be rapidly deployed—was "the biggest thing going." As the largest software project to date, with a $100 million budget, Site Defense offered unprecedented resources to develop and expand a managerial vision of software engineering. [57]

The project managers, William Besser and Robert Williams, decided to invest in automated "tools"—pieces of software designed to improve the programming process. Such project management software was not new. As Harvey Sapolsky and others have argued, Polaris managers used similar software to gain an air of authority, persuading their critics that everything was under control.[58]

In the 1970s, many managers began promoting such tools as essential to software engineering. For example, Williams described software development tools for requirements engineering, software design, implementation, and testing, arguing that they were part of a "definitive and effective reliability-oriented software production methodology."[59] Software developers of more limited means could not necessarily use software tools—the programs crashed all but the fastest of machines.[60] But tools were characteristic of a managerial approach to software engineering, one embraced by other large software developers.

The sheer size of such projects gave them a remarkably large influence on an emergent field of software engineering. For example, in the 1970s, Boehm and others at TRW gathered software data not only for the Army, but also for the National Bureau of Standards and the Air Force, eventually publishing the results in the TRW Series of Software Technology.[61] Many government agencies institutionalized software research and development: in 1975, the Air Force established its Data & Analysis Center for Software (DACS).[62] The same year, Boehm helped start a new conference series that would be devoted to software engineering.

2   The Revival of Software Engineering

2.1   Managing Chaos

In 1974, Boehm joined with two computer science professors—Raymond Yeh at the University of Texas Austin, and Tony Hoare at Queens University in Belfast—to organize an International Conference on Reliable Software (ICRS).[63] They aimed to internationalize what had begun with the Institute of Electrical & Electronics Engineers (IEEE) Symposium on Reliable Software in 1973, an effort to tackle unreliability "from a combined practical and theoretical viewpoint."[64] Sponsored by the IEEE and more academically oriented Association for Computing Machinery (ACM), the ICRS drew academic and industrial researchers from around the world to Los Angeles, California, in the spring of 1975. Several U.S. participants were inspired to rapidly organize the National Conference on Software Engineering (NCSE), held in September 1975, with the sponsorship of the recently formed IEEE Computer Society and the U.S. National Bureau of Standards. Boehm described the NCSE as "sort of a quick and dirty thing," but noted that it became "a big deal" in 1976, as the International Conference on Software Engineering (ICSE).[65] Regular meetings of the ICSE have continued since then, with the sponsorship of the ACM and IEEE Computer Society.[66]

Although ICSE was deliberately diverse, it reflected the managerial interests of very large software developers. Significantly, the top four companies represented at ICSE 1976—TRW, Bell Labs, SofTech, and Computing Sciences Corporation—had worked on, or were then working on, missile defense software projects. Furthermore, over 10 percent of the presenters at ICSE 1976 were associated with the ballistic missile defense software project. These presenters were more than three times as likely as other presenters to emphasize management tools. [67]

Like the NATO-sponsored Software Engineering Conferences, ICSE included participants from academia, industry, and government, working primarily in the United States and Europe. But the two conference series shared little else. Whereas the organizers of the late 1960s conferences invited papers, the organizers of ICSE put out open calls for papers, with acceptance rates hovering around 20 percent or less.[68] Out of 57 hand-picked participants in the 1968 conference, one-third returned for the 1969 meeting. By contrast, only three of the 1968 conference participants (4 percent) attended the 1976 International Conference on Software Engineering, and only five joined any ICSE meetings in the 1970s. The ICSE gatherings thus represented a new community, and became what Boehm has termed the "primary unifying" activity in software engineering.[69]

Enthusiasts for software engineering did not stop with conferences. In 1976, Yeh became the founding coeditor of a new peer-reviewed journal, *Transactions on Software Engineering*, published by the IEEE Computer Society. Meanwhile, the program chair for the NCSE, Thomas Steel, became the chair of a new ACM Special Interest Committee on Software Engineering (SICSOFT). By 1977, when SICSOFT became the more permanent Special Interest Group, SIGSOFT, it boasted more than 4,000 members, making it the third largest special interest group within the ACM.[70] As SIGSOFT's new chairman, Wasserman observed: "Few people track software engineering back to Garmisch in 1968."[71]

But just what was software engineering? In the inaugural volume of SIGSOFT's newsletter, *Software Engineering Notes (SEN)*, Steel attempted to put the field "in perspective" with a short anecdote about a priority dispute:

Surgeon (confidently):   I need only quote Genesis II, 21–22, to prove that the surgeon's art was prerequisite to that which is traditionally viewed as the world's oldest profession. "And the Lord God caused a deep sleep to fall upon Adam, and he slept; and He took one of his ribs, and closed up the flesh instead thereof. . . ."

Engineer (patronizingly):   You forget, sir, that according to Genesis I, 1, "In the beginning God created the heaven and the earth." Certainly, constructing the whole universe out of chaos is a feat of engineering.

Programmer (triumphantly):   Who do you suppose created chaos?

As Steel explained, the aim of SICSOFT was "to reduce the chaos in every way possible."[72]

Indeed, a common concern with complexity was perhaps the only theme that united a broad SIGSOFT membership. Peter Neumann, a mathematician working on provably secure operating systems at Stanford Research Institute and the editor of *SEN*, refused to set "explicit boundaries" on acceptable contributions, noting that topics such as "Software Engineering,

Photosynthesis, and Religion" were "neither suggested nor excluded." He welcomed "good technical controversy," in hopes that the new group might "make real the use of the word 'engineering' in software engineering."[73]

Software engineering remained diverse. In 1978, the most-requested topic by readers of *SEN* was "design methodology, followed by programming methodology," but the list went on:

Also frequently mentioned were tools, then "practical experience" and management topics. Also worthy of note were proofs, psychological and philosophical aspects of software engineering, software metrics, reliability, and security. Several readers wanted more on interfaces, e.g., between theory and practice, between "computer science" and "software engineering," and between technical and management issues.[74]

2.2   Languages as a Managerial Tool

Management interests drove much of the funding for software engineering—a fact that worried many researchers. For example, managerial interests drove the development of a standard programming language for military computers in the 1970s. Air Force colonel William Whitaker began the project as the Defense Department's first initiative to "Reduce the High Cost of Software." Trained as a nuclear physicist, Whitaker grew familiar with the challenges of nonstandard programming languages in the 1960s, while overseeing scientific computation for the Air Force Weapons Laboratory. In 1973, a few months after Whitaker joined the Electronics and Physical Sciences Section in the Directorate of Defense Research & Engineering (DDR&E), he attended a Symposium on the High Cost of Software, at the Naval Postgraduate School in Monterey. There he heard Boehm present the CCIP-85 study's estimates of growing software expenses. Whitaker recalled that the findings were so "shocking" they were "rejected out-of-hand." Whitaker went on to initiate a new study by the Institute for Defense Analyses, as validation of Boehm's results.[75]

With numbers in hand, Whitaker began making the case for a new, general-purpose language for the Defense Department. He targeted "embedded computer systems"—those used in military operations, rather than those used in offices to handle payroll and other routine administrative tasks.[76] Despite initial opposition, Whitaker eventually succeeded in establishing a Higher Order Language Working Group (HOLWG) that included representatives from each of the three services.[77] The group began defining language requirements with programming specification entitled "Strawman," and refined the requirements through several successive specifications: Woodenman, Tinman, Ironman, and Steelman.The final specifications were sent out to software contractors. By May 1979, when Honeywell-Bull won a contract to develop the language, the language had been dubbed "Ada," in honor of Augusta Ada Lovelace, Charles Babbage's "programmer."[78]

Managerial interests shaped the design of Ada from the beginning. Of six goals for the new language, one was primary: to "facilitate the reduction of software costs."[79] This goal demanded "transportability," "flexibility," "efficiency," long-term "readability," and even "reliability."[80] Whitaker recalled that such "ilities" sounded good but did "not lend themselves to a quantifiable or rational assessment of programming languages."[81] Indeed, one of the software developers, John Goodenough, noted: "Certainly, the Navy felt that only the Navy could design a language that would meet Navy needs. Only the Air Force could design a language that would meet Air Force needs. Only the Army for Army needs. Overcoming that kind of thinking was a major bureaucratic in-fighting win. . . ."[82]

Resistance also emerged from researchers who opposed the project's managerial agenda. As a reviewer of the Strawman requirements, Dijkstra objected that their focus on cost savings could have "catastrophic consequences."[83] He argued that higher quality should be the main goal of language design, and cost reductions only a "fringe benefit." [84]

Researchers also objected to managerial notions that languages were tools for enforcing good programming habits, irrespective of the skills of software designers. Indeed, many languages were designed to enforce programming rules. For example, "strongly typed" languages were developed to eliminate typographical errors from final programs, by checking for type consistency before finalizing an executable program.[85] While features such as strong typing came to be widely accepted, some software engineers objected to an overemphasis on such features.

For example, at a 1977 conference on language design, one panel devoted itself to discussing the *limits* of languages. There David Parnas, a young professor of computer science at the University of North Carolina, presented a paper on "BLOWHARD," or "Basic Language—Omnificent with Hardly Any Research or Development." BLOWHARD was a farce—it offered programmers absolutely no options—proposed only to highlight the dangers of an "alternative farce," an overly elaborate and complicated language.[86] Although Parnas was not opposed to new languages—he had earned his PhD under Alan Perlis, a leader in the ALGOL development—he argued that "improvements in software reliability are best obtained by studying software design not by designing new languages."[87] Peter Neumann noted: "a good designer can produce a good product with all but the worst tools.

A bad designer can produce a bad product with the best of tools."[88] Indeed, Parnas's foundational work articulated design principles that would enable software to be changed and adapted more readily, thereby reducing errors and the consequent need for debugging.[89]

Dispute over the value of languages to good software design partly reflected the establishment of software engineering as a field, distinct from that of programming languages.[90] But it also reflected divisions within the new field of software engineering, between researchers who emphasized innovative thinking and correctness, and managers who focused on standardization and cost cutting. Dijkstra blamed American management philosophy, which "aimed at making companies as independent as possible of the competence of their employees."[91] He would "shudder" at meeting Boehm and his associates working on missile defense software for the Army, because they used "jargon" such as "human factors" and "human engineering."[92] He felt that their emphasis on improving programming tools only furthered a false charter for software engineering: "How to program if you cannot."[93]

2.3   Security and Formal Proof

Managers concerned about cutting costs also drew upon efforts to formally prove the correctness of programs. One reason the Defense Department duplicated computer resources was that there was no way to secure data on computers that could be accessed by multiple users—individuals with a low level of clearance could easily get access to data at a higher level of clearance. In 1967, Willis Ware, a computer analyst at RAND, and chair of the National Security Agency's Scientific Advisory Board, led a Defense Science Board task force in a study of the security problem.[94] In 1970, the task force concluded that since it was "impossible to address the multitude of details that will arise in the design or operation of a particular resource-sharing computer system," security problems must "be solved on a case-by-case basis."[95] Ware's group recommended that the Defense Department fund several areas of research, but prompted little concerted action.

A second study, commissioned in 1972 by the Air Force, made a much stronger impact. Directed by an independent consultant, James P. Anderson, the Computing Security Planning Study distinguished itself from Ware's in two key respects. First, whereas Ware had outlined a piecemeal research agenda, Anderson's group emphasized the need for a "comprehensive and coordinated attack" on the security problem, including designing security "into a system from its inception."[96] The report also recommended four related research programs totaling nearly $30 million, and took special note of academic work by Dijkstra and others, which aimed at developing programs that could be formally proven correct.[97] Second, whereas Ware had expressed concern about the *expense* of securing computers, Anderson focused on the potential for cost *savings*. Air Force duplication was costing approximately $100 million a year.[98]

Anderson's study persuaded both the National Security Agency and ARPA to fund significant research into formal verification throughout the 1970s.[99] But while verification offered a seductive vision for managers—guaranteed reliability—many software experts emphasized its limits. At the International Conference on Reliable Software, Harlan Mills noted that while programmers might prove that a program met its formal specifications, they could not prove that the program would do what it *should* do. Furthermore, he concluded: "there is no foolproof way to prove that a program is correct. The fundamental difficulty is not in programming, but in mathematics . . . a human activity subject to human fallibility. It has no basic secrets of truth or reason."[100] Mills admonished programmers not to go "looking for some schoolboy magic to replace your own responsibility."[101]

Perlis and two younger computer scientists—Richard Lipton and Richard DeMillo—made similar points at a conference two years later, arguing that mathematics was reliable to the extent that it was subject to human review and critique. However, they further argued that program proving was a less social, and therefore less reliable, form of mathematics. "The proofs of even very simple programs run into dozens of printed pages," and thus were unlikely to be "read, refereed, discussed in meetings . . . or used."[102] They claimed that "the kind of mathematics that goes on in proofs of correctness is not very good," and warned: "Program proving is bound to fail in its primary purpose: to dramatically increase one's confidence in the correct functioning of a particular piece of software."[103]

When Dijkstra read these words, he dubbed the paper "A Political Pamphlet from the Middle Ages." He objected to the premise that program "proofs are not communicated and subjected to the judgment of others" and insisted that "the correct functioning of particular pieces of software" is the subject of a lively interchange of experiences between scientists active in the field.[104]

This exchange sparked a lengthy debate in the pages of *SEN*. Several individuals noted that efforts to verify programs had proven useful in real-world applications. For example, a manager at Chase Manhattan Bank noted that efforts to verify programs had increased confidence in "real-world programs." Significantly, he emphasized the value of formal verification for cutting costs: "errors propagate errors, and . . . an error prevented for $10 can save $50 to $90 in the next development phase."[105] As this suggests, work on program proving was often justifiable only in terms of managerial interests—a fact that many researchers detested.

In practice, lofty standards of proof were lowered by the formidable complexity of real-world systems. For example, in 1973, Neumann began leading an ARPA-sponsored effort to develop a Provably Secure Operating System (PSOS) at SRI. Seven years later, the final project report acknowledged that PSOS was only a "Potentially Secure Operating System," noting: "PSOS has been carefully designed in such a way that it might someday have both its design and its implementation subjected to rigorous proof."[106] Similarly, an effort at the University of Texas in Austin to develop a provably secure military computer network as part of the WMMCCS, Autodin II, was canceled in 1982 in favor of encryption.[107]

3   Computer-Related Risks

In the late 1970s, a new theme emerged in the pages of *SEN*: learning from failure. In November 1978, *SEN* was devoted to a Software Quality and Assurance Workshop, where Robert Glass, a computing consultant and industry writer, presented a paper on "Computing Failure." Glass noted that while "most professional papers" focused on success, "some of our most lasting learning experiences are based on failure."[108] Glass went on to publish books about fiascos and conflicts in software engineering.[109]

While Glass highlighted corporate disasters, others highlighted the risks that computers might pose to public safety. In March 1979, Neumann reported some "SENsitive NEWS ITEMS," including breaches in computer systems at banks and the discovery of a mistake in software used to design nuclear reactors. The design error prompted the shutdown of five nuclear plants, while engineers sought to correct the problems. Neumann highlighted problems not only with program bugs and inaccurate physical models, but also with "the intrinsic gullibility of the computer illiterate—including many administrators and engineers as well as laypersons."[110] He concluded:

As I am writing this, a radiation leak has just occurred at the Three Mile Island nuclear plant #2 near Harrisburg, PA. Although probably not computer-related, this accident resulted from operator error and multiple faults. . . . This further example of Murphy's Law should be taken seriously by purveyors and users of "reliable" computer systems.[111]

The Three Mile Island meltdown occurred on March 28, 1979, and Neumann's editorial went to press in April. In the next issue of *SEN*, Neumann forwarded a call for "true stories on systems (mis)behavior, e.g., that 'surprises, baffles, spoofs, or even hoodwinks its designers and keepers,'" which would be discussed at an upcoming symposium on Operating Systems Principles.[112]

Letters flooded in. For example, James Horning, a computer scientist at Xerox's Palo Alto Research Center who was working on formal verification, worried that the software industry did not provide sufficient incentives to make software reliable. He predicted that within a decade "there will be several major disasters that are directly (and properly) attributed to faults in computer programs: airplane crashes, or nuclear accidents, or bridge collapses, or. . ."[113] Horning described "a recurrent nightmare in which one of these disasters has led to a court case." The prosecutor accused the program developer of negligence. But the defendant accused the user of foolishly relying on a large program: "no large program is fault-free." Horning confronted three choices: should he testify for the prosecutor, the defendant, or "chicken out and refuse to testify"? Delivering these remarks before an audience of about 200 people, Horning asked what they would do. Approximately one-fourth indicated they would testify for the prosecution, about one-third would testify for the defense, but nearly half would chicken out.[114]

Horning argued that "most of those who actually pay for computer programs . . . do not care significantly about their reliability," and that "until they do, improvements in reliability will merely be "side effects" of efforts to improve things that they do care about, such as the cost of programming." He challenged readers to find ways of increasing the importance of reliability to "those who pay the bills."[115]

In the months and years following the Three Mile Island meltdown, *SEN* began to publish news of computer-related disasters. By 1983, Neumann recommended that somebody "write a book that collects and analyzes the most interesting cases of computer-related problems, accidents, disasters, penetrations, misuses, etc., documenting them carefully," so that historians and systems designers could learn from failure. He recommended posting an advertisement in "computer and literary magazines":

Author needed. Must be dedicated researcher conversant with computer technology, committed to the importance of human life, willing to follow endless leads, rumors, and apocrypha, persistent enough to pursue dark secrets. Many publishers interested. Best-seller possibilities. Potential subtitle: Bugs, Plugs, Slugs, and Thugs.[116]

Twelve years later, Neumann himself would write the book.[117]

Concerns about risks to the public were partly driven by the growing prevalence of microprocessors. In the 1960s, computers were large machines, found primarily in workplaces—offices, laboratories, or military operations. But in the 1970s, microprocessors began to find their way into pacemakers, watches, microwaves, and myriad other popular electronics. While the packaged software industry had matured somewhat, microprocessor programming was a field full of amateurs. It was out of this field that the personal computer would eventually develop.[118]

The microprocessor industry led William McKeeman, a computer science professor at the University of California Santa Cruz, to worry about two conflicting trends. First, the rapid expansion of the industry was "pressing into service a large number of people without much formal training." The average skill of programmers was "dropping, and along with it the quality of software." At the same time, "the user community is expanding to approach the entire literate population with programs running microwave ovens, automobile engines, hand calculators, telephones, airplanes, banks, gas pumps and heart patients." This was "a very inelastic audience. Nearly any problem is beyond the skill of the user to understand, or work around, much less to repair."[119] Similarly, *SEN* reported many risks related to microprocessors, which could cause faults in anything from pacemakers to transportation and power systems.[120]

SEN's growing concern with risk also reflected broader shifts in public attitudes toward large and complex technological systems. The energy crisis of the 1970s seemed to confirm the basic tenets of the "appropriate technology movement"—that large, complex technological systems concentrated power in the hands of a few elites who could not be trusted.[121] The pages of *SEN* articulated many concerns that were shared by sociologists of risk and organizations. For example, Yale sociologist Charles Perrow published a groundbreaking book analyzing the Three Mile Island meltdown as a "normal accident." Perrow argued that failures were inevitable in any system with complex, tightly coupled interactions, because unexpected interactions would emerge and evolve too rapidly for human organizations to respond properly. He also highlighted the problems of reforming organizations that were driven primarily by economic incentives—a point that was all too familiar to many in the software engineering community.[122]

Thus, by the early 1980s, a common concern with managing risk in complex systems had come to center the field of "software engineering." To be sure, the very notion of "software engineering" would remain controversial and divisive. Nonetheless, the establishment of new forums for discussing the challenges of complex software turned past problems into something that was not only "well-remembered," but "well-documented." It also established a community of computer workers who followed closely as the United States invested in increasingly complex and automated systems for controlling nuclear weapons. As we will see in the next chapter, these individuals began to speak up, even before President Ronald Reagan proposed the most ambitious system ever: a missile shield that would render nuclear weapons "impotent and obsolete."

8   Nature and Technology in the Star Wars Debate

In March 1983, the President asked us, as members of the scientific community, to provide the means of rendering nuclear weapons impotent and obsolete. I believe that it is our duty, as scientists and engineers, to reply that we have no technological magic that will accomplish that.

—David Parnas, June 28, 1985

In June 1985, David Parnas resigned from a computing panel for the Strategic Defense Initiative, or "Star Wars" missile defense program. Based on more than twenty years of experience in software research and development, he warned that complex software was prone to unreliability for "fundamental" mathematical reasons that would "not disappear with improved technology."[1] Parnas's declaration struck at the heart of Star Wars, a program that insisted high technology could do almost anything—even make nuclear weapons "impotent and obsolete." It also set off a firestorm of controversy among computer experts, who remained as committed as ever to technological progress, but increasingly questioned whether complex software might confront fundamental limits.

The goals of Star Wars were ambitious, and the stakes were high. Space-based missile defense would require highly automated decision making by the most complex software system ever developed, yet system testing would be partial and incomplete until the moment of truth. Experience with increasingly complex nuclear command and control systems was not promising. And while space-based defenses remained hypothetical, far less capable space weapons raised fears of an imminent nuclear war, at a time when the United States and Soviet Union both possessed the most potent nuclear arsenals in history.

In this context, Parnas's claims of fundamental limits to software fueled and reshaped an ongoing debate. Would high-tech defenses increase or decrease security? What were the risks of placing confidence in such a complex technological system? And were there really "fundamental" limits to the reliability of complex software?

Most accounts of the Star Wars controversy focus on elite physicists and policymakers. By contrast, this chapter shows how the field of software engineering began to shift the terms of a decades-old debate about defense. To be sure, physicists remained the most visible experts on missile defense. Hans Bethe, Richard Garwin, and others renewed their analysis of the race between offensive countermeasures and defensive technologies, critiquing defensive proposals from the nuclear weapons laboratories, industry engineers, and military space enthusiasts. While optimistically assuming that the defensive weapons could be developed and work perfectly, these physicists showed that the effectiveness of even the most promising technologies would face absolute limits: the laws of nature.

As computer experts reflected on decades of experience with software engineering, many concluded that their technologies were confronting similar limits. Using the disciplinary repertoire established in the 1970s— quantitative rules about the economics of complex software, as well as a qualitative record of risks and faults in complex systems—they sought to construct persuasive arguments about the risks of a missile defense system. By underscoring "fundamental" or "essential" limits to the reliability of software, many software engineers fashioned their arguments after those of physicists. While remaining optimistic about technological progress, they agreed that software engineering would be forever constrained by the "irreducible essence" of complex computer programs. Furthermore, arguments about the limits of complex technology raised a more profound question: what did it mean to produce a "trustworthy" or "reliable" system that was designed to be used only once?

To address this question, we turn first to the early 1980s, an era of weapons and policies that moved steadily toward what Herbert York dubbed the "ultimate absurdity."[2]

1   Risking Disaster, Promising Progress

1.1   Headless Horseman of the Apocalypse

When President Ronald Reagan took office in January 1981, he inherited a controversy over the MX missile. Capable of delivering nuclear devastation with unprecedented speed and accuracy, MX would carry 10 independently targetable warheads, each three times more accurate than existing U.S. warheads, and 17 times more destructive than the bomb that destroyed

Hiroshima. Reagan dubbed MX the "peacekeeper," to imply it would be a powerful deterrent to war. But there was one hitch: the MX might so undermine the Soviet retaliatory force that Soviet leaders would fear a U.S. first strike, and be tempted to strike preemptively. It was the problem of "crisis stability" identified by Thomas Schelling two decades earlier.

MX supporters sought to avoid the problem by ensuring that the deployment would be able to survive a first strike, making preemption completely futile. Most strategists sought to keep MX moving, whether on balloons or barges, in tunnels or trains. But despite 30 different deployment proposals, generated during a decade of analysis, not one could survive combined critiques by scientists, military planners, and citizens (who typically cared less about survivability than about ensuring that nuclear missiles would be "not in my backyard"). When Reagan inherited the controversy in 1981, he proposed to temporarily base the MX in a "dense pack" of super-hardened silos in Arkansas, while studying more survivable options.[3]

Physicists such as Hans Bethe objected that such a non-survivable deployment, however temporary, would only increase the risk of war.[4] And indeed, as systems such as MX shortened decision-making times, Soviet military leaders began to warn that they would be forced to launch a retaliatory strike more rapidly, even in the face of ambiguous evidence.[5] Meanwhile, rising tensions left the Soviet military increasingly wary of U.S. aggression. Arms control negotiations ground to a halt with the Soviet invasion of Afghanistan in 1979. As NATO prepared to deploy Pershing II intermediate-range missiles in West Germany in the fall of 1983, weapons that would only offer the Soviets six minutes of warning, Soviet leaders warned that the deployment would force a faster and more reactive launch policy. Indeed, on September 1, 1983, Soviet air defense officers shot down Korean Air Flight 007 after it strayed into Soviet territory. At first Soviet leaders unrepentantly declared that the airline was spying, but officials slowly acknowledged that the air defense officers were mistaken, explaining that high tension had made them "trigger-happy."[6]

The United States was moving toward similarly trigger-happy policies. The most economical way to ensure the survivability of MX would be to adopt a launch-under-attack policy, firing nuclear missiles after "a high-confidence determination" that a Soviet attack was under way, rather than waiting to determine the extent of the attack, or whether it might be an accident.[7] Accordingly, the Reagan administration aimed to develop Command, Control, and Communications (C^3) systems that could be trusted with a launch-under-attack decision.[8] In fact, Secretary of Defense Caspar Weinberger made C^3 the Pentagon's "highest priority element" in its strategic modernization program, because it was critical to the administration's declared nuclear policy: to "prevail" in the event of a nuclear war.[9]

However, strategic analysts objected that the improvements would not likely enable commanders to maintain control over even the most limited nuclear exchange, let alone to "prevail." Physicists emphasized that the electronics and computers linking various command and control systems would likely be destroyed by the electromagnetic pulses from nuclear blasts.[10]

Furthermore, the organizational challenges associated with complex computers began to draw more attention. For example, on the morning of November 9, 1979, officers at NORAD saw warning of a nuclear attack and launched Air Force interceptors while raising the alert on the Strategic Air Command. Fortunately, the alerts were stopped after six minutes, when officers could not find independent confirmation of the attack. The false alarm was blamed on a simulation tape that was being run on an upgraded, soon-to-be-installed computer, but analysts never could explain how the simulated signals showed up in the online NORAD computer. Less than six months later, in the early morning hours of June 3, 1980, officers in several command posts again saw signals indicating a nuclear attack and once again prepared for the worst. For nearly an hour, nuclear forces alternated between various stages of alert while commanders attempted to make sense of strange and inconsistent warnings. The alert was eventually called off, but just days later, a similar set of events recurred. It turned out that a computer chip had malfunctioned.[11]

In July 1980, the editor of *Software Engineering Notes*, Peter Neumann, highlighted both failures in what had become an ongoing conversation about computer-related risks. He warned readers to be wary of "any system that claims to be reliable, secure, etc., in an environment in which anything important really depends upon those claims."[12] Neumann and others studying safety-critical software drew upon analyses by social scientists, which emphasized the organizational challenges of command and control—captured by the title political scientist Paul Bracken chose: "Headless Horseman of the Apocalypse." Computer experts emphasized that poorly designed, obsolete, or fallible software easily exacerbated the organizational challenges of command and control, such as fragmentation and crossing lines of authority.[13]

Meanwhile, physicists were less concerned about the fallibility of software than they were about a growing U.S.-Soviet race for antisatellite weapons (ASATs), which could threaten the delicate "eyes" of nuclear command and control systems. The Soviets led the ASAT charge, with highly visible tests in the late 1970s and early 1980s. Only about half of the Soviet tests were successful, and none could go beyond low orbits to target more distant military satellites. Nonetheless, the Reagan administration cited the Soviet effort when it launched an effort to deploy ASATs within five years.[14]

Arms controllers warned that ASATs would contribute to crisis instability. If the Soviets feared an American ASAT capability, an "ordinary" satellite failure might suddenly seem to bode a surprise attack. Similarly, if the Soviets possessed ASATs, American military leaders would suddenly get nervous and trigger-happy whenever they lost satellite signals. Thus, in 1983, Hans Bethe, Richard Garwin, and other elite physicists working under the auspices of the Union of Concerned Scientists, drafted a treaty to ban weapons in space. With the signatures of 17 prominent defense experts, they sent the petition to U.S. and Soviet leaders, and presented it to the Senate Foreign Relations Committee. Soon a bipartisan group of senators called upon Reagan to negotiate a ban to space weapons—but to no effect.[15]

Meanwhile, the Reagan administration's aggressive rhetoric left many arms controllers worried that high-level officials might be all too willing to start a nuclear war. Reagan largely continued weapons and plans that had been in development for decades, but his declared policy of "prevailing" in a nuclear war was far more controversial.[16] Did the administration really think nuclear war was winnable? In the face of public critique, Defense Secretary Weinberger insisted: "You show me a Secretary of Defense who's planning not to prevail, and I'll show you a Secretary of Defense who ought to be impeached."[17]

Even more controversial were a series of high-level comments suggesting blithe ignorance inside the administration about the effects of nuclear war. Most infamously, Thomas K. Jones, Reagan's deputy under secretary of defense for research and engineering, strategic and nuclear forces, told reporter Robert Scheer that the United States could recover from nuclear war with the Soviet Union in only two to four years. Jones explained that the millions of people who relocated from cities to the country would simply dig holes, cover them with doors and about three feet of dirt, and thus avoid fallout. He explained: "It's the dirt that does it. . . . If there are enough shovels to go around, everybody's going to make it."[18]

1.2   Accidental Nuclear War?

Fueled by the Reagan administration's brash nuclear talk, public interest science groups renewed efforts at arms control in the early 1980s, mobilizing a broad base of support. For example, Helen Caldicott, who cofounded the Physicians for Social Responsibility to oppose nuclear testing in the early 1960s, and then opposed nuclear power in the 1970s, shifted back to weapons in the 1980s. With the rise of the feminist movement, she also broadened the base of opposition, founding Women's Action for Nuclear Disarmament in 1982. Religious groups such as the American Friends Service Committee (AFSC) also renewed disarmament efforts in the early 1980s.[19]

Public interest science and religious groups came together through the leadership of Randall Forsberg, a long-time student of arms control and MIT-trained political scientist. In 1980, Forsberg drafted a Call to Halt the Nuclear Arms Race, proposing that the United States and Soviet Union agree to a mutual freeze on all nuclear weapons testing and production. Forsberg worked with MIT physicists and the AFSC to disseminate his Call. In March 1981, Forsberg drew leaders from a wide range of peace groups to a conference to discuss strategies for unifying and growing their campaign. Leaders of the nuclear freeze brought human suffering into discussions of nuclear war, and rendered it palpable with films such as *The Day After*, which aired on the ABC network in November 1983.[20]

Many computer experts felt they had a new role to play in the disarmament movement. In 1982, several researchers at Xerox Palo Alto Research Center began to discuss concerns that nuclear war was imminent. They invited local researchers, such as Peter Neumann at Stanford Research Institute, and a group of computer workers from throughout Silicon Valley, to form a new group: Computer Professionals for Social Responsibility (CPSR).[21]

Members of CPSR aimed to raise awareness about the risks associated with the computerization of war. Alan Borning, a CPSR cofounder and recent graduate of Stanford's computer science department, began to draw on the many complex systems failures that Neumann had helped document in *Software Engineering Notes*, as he analyzed the ways that complex computer systems might spark an accidental nuclear war. He argued that the U.S. "launch-under-attack" scheme placed far too much faith in complex computerized systems, and eventually published an article explicitly opposing the recommendations of physicists such as Richard Garwin, who argued that launch-under-attack could be an effective and reasonably reliable way to enhance deterrence.[22] Reflecting on Borning's analysis, one CPSR member noted that just as physicists bore responsibility for advising the government about the risks of nuclear weapons, computer scientists "bear the responsibility of advising the government of the dangers of today's weapons systems," because "the heart of today's weapons systems is the computer."[23]

In October 1982, CPSR members urged the ACM Council to take a stance on the relationship between computer technology and the risk of nuclear war. As in the 1960s, the ACM refused to make such a potentially "political" statement. But the council chartered a study that was eventually completed by an "ad hoc Committee on Computer Systems Reliability." The seven-member committee was chaired by University of Michigan professor Aaron Finerman, who had agreed with McCracken in the late 1960s that the ACM was "long overdue in warning against over-reliance on any computer-driven fail-safe system."[24] The group also included several members of CPSR, such as Anthony Ralston, a sponsor of the computer professionals against ABM in the late 1960s, and James Horning, whose research on safety-critical computing had given him a "recurrent nightmare" of a software-related disaster.[25]

In July 1984, the committee proposed that the ACM pass a resolution declaring the "reality" that "computer systems can and do fail," sometimes posing "extreme risk to the public."[26] They warned: "Increasingly, human lives depend upon the reliable operation of systems such as air traffic and high-speed ground transportation control systems, military weapons delivery and defense systems, and health care delivery and diagnostic systems."[27] The ACM Council approved the resolution, and tasked the Committee on Computers and Public Policy (CCPP) with a study of computer system reliability and risks.[28]

Peter Neumann, the chair of the CCPP, began responding to his new charge by establishing a new section of *SEN*, entitled "Risks to the Public in Computer Systems." He also began moderating an e-mail forum, "RISKS." RISKS covered a broad range of subjects, and frequently included discussions of the dangers of complex weapons systems.

Arguments about computing were well received in a broader anti-nuclear movement that combined expert knowledge with common sense. As computers became increasingly familiar in everyday life, Americans were growing more familiar with their fallibility. Indeed, the film *Wargames* (1983), in which a teenage boy hacks into a Defense Department computer and sets off a chain of events that nearly lead to nuclear war, was perhaps the most popular anti-nuclear movie of the era.

1.3 "A Vision Which Offers Hope"

By the early 1980s, a broad-based anti-nuclear movement posed a serious threat to Reagan's policies. In March 1982, both the Senate and the House introduced legislation calling for a freeze on all further weapons production. Though freeze legislation was defeated in both houses of Congress, in November 1982 voters approved resolutions supporting a nuclear freeze in eight states and several major cities that collectively contained nearly one-quarter of the nation's population.[29] The majority of those who had the opportunity to vote clearly supported the freeze.[30] Reagan's proposed MX deployment was already in trouble—even the Air Force and Joint Chiefs of Staff opposed the dense-pack scheme. Bolstered by growing anti-nuclear sentiment, the U.S. Senate easily rejected the MX deployment in December 1982.[31]

Seeking to ease public anxieties and gain support, Reagan's staff turned to Ballistic Missile Defense (BMD). The Army was already arguing that its Low Altitude Defense System (LoADS), a relatively cheap radar and computer that controlled a nuclear-tipped interceptor, could help defend MX from a first strike, contributing to a survivable deployment. As part of its "Homing Overlay Experiment" (HOE), the Army was also developing a "hit-to-kill" vehicle that would be able to get close enough to an incoming warhead to destroy it without nuclear weapons, potentially contributing to a thin defense of populations.[32] While LoADS emerged from the "Safeguard" debate and recommendations for more rapidly deployable defenses, the HOE emerged from objections to nuclear-tipped interceptors around cities in the "Sentinel" debate.

However, Reagan's advisers were less interested in the Army's proposed defenses, than in far more ambitious proposals, to provide a nearly perfect protection of populations. These more futuristic visions of defense emerged not from the Defense Department, but from a disparate set of technologists, industrialists, and policymakers. For example, Wyoming senator Malcolm Wallop and one of his staff members, Angelo Codevilla, took up the proposal of Maxwell Hunter, an engineer at Lockheed Missiles and Space Company, to deploy laser weapons in space. Separately, Daniel Graham, a retired Defense Intelligence Agency official, proposed using space-based kinetic kill vehicles to destroy missiles. (This proposal resembled the Air Force's Ballistic Missile Boost Intercept [BAMBI] scheme, which had been rejected at ARPA in the late 1950s.) Still another proposal emerged from Lawrence Livermore Laboratories, where physicists Edward Teller and Lowell Wood were advocating the rapid development and deployment of a nuclear-powered X-ray laser.[33]

Program managers in the Defense Department and at Livermore fought the advocacy of Wallop, Graham, and Teller. They argued that reorienting research toward deployment would only waste resources. The energy of lasers would need to grow by orders of magnitude before they could destroy missiles, while sensors would need to improve in order to detect and track objects precisely. Arms controllers also objected that since space-based defenses would inevitably threaten satellites, and would likely be vulnerable to attack, they would make a tempting target for a first strike, fueling dangerous instabilities.[34]

Although Defense Department authorities resisted pressures, enthusiasts for space-based weapons helped produce an appealing vision of a more peaceful, defense-dominated world. For example, Wallop and Codevilla published an article in *Strategic Review*, claiming: "Several dozen laser weapons systems deployed in space would revolutionize the strategic equation as we have known it for nearly two decades—above all by decisively tipping the balance of modern warfare in favor of the defense and radically mitigating the potential destructive effects of war."[35] Teller's promise that X-ray lasers in space could stop a "mass missile attack" made the front page of the *New York Times*.[36] Journalists, seeking to make sense of it all, referred to the film *Star Wars*, and used graphics that depicted lasers precisely destroying objects in space, reifying technology that was far from proven (figure 8.1).[37]

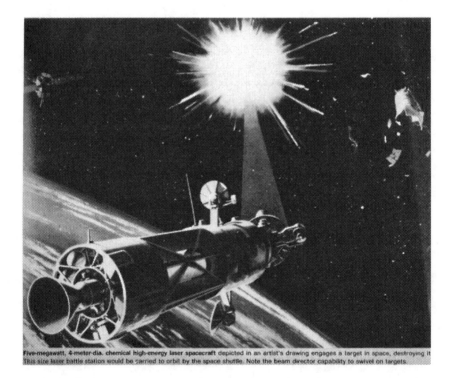

Five-megawatt, 4-meter-dia. chemical high-energy laser spacecraft depicted in an artist's drawing engages a target in space, destroying it. This size laser battle station would be carried to orbit by the space shuttle. Note the beam director capability to swivel on targets.

Figure 8.1
An artist's concept of how laser weapons would destroy nuclear weapons cleanly and precisely.
Source: "Laser Technology Demonstration Proposed," *Aviation Week and Space Technology*, February 16, 1981, 16. (The image is presumed to be in the public domain.)

Facing opposition from a broad-based anti-nuclear movement, Reagan's advisors began mobilizing the advocates and rhetoric of these revolutionary defenses. National security adviser William Clark and his deputy Robert McFarlane hastily and secretly drafted an addendum to Reagan's speech on the military budget.[38] On March 23, 1983, Reagan delivered a carefully staged, nationwide televised address from the White House, where nuclear physicists such as Edward Teller and Hans Bethe had been invited for a special dinner and briefings. After justifying a continued escalation of military spending on offensive weapons such as MX, Reagan changed his tone. "Wouldn't it be better to save lives than to avenge them?"[39] He continued with a vision that drew deeply on long-established American narratives of technological progress:

Let us turn to the very strengths in technology that spawned our great industrial base and that have given us the quality of life we enjoy today. . . .

I call upon the scientific community in our country, those who gave us nuclear weapons, to turn their great talents now to the cause of mankind and world peace: to give us the means of rendering these nuclear weapons impotent and obsolete.[40]

Reagan did not mention lasers or space-based weapons in his speech, but White House officials confirmed that his proposal might include such technologies. Policymakers and journalists promptly dubbed his vision "Star Wars," and the label stuck.[41]

Many physicists objected. In a *New York Times* op-ed piece, Garwin outlined a host of countermeasures that could easily overwhelm defenses, called again for a ban on space weapons, and emphasized that American satellites were "too valuable to risk for a will-o'-the-wisp far more likely to lead to increased Soviet offensive forces than to a reduction of the threat to our lives."[42] But Garwin's article appeared next to Teller's promises that "a wide range of good and ingenious technical plans" might enable the "transition from mutual assured destruction to mutual assured survival."[43] Longtime promoters of exotic defenses consistently supported Reagan's vision, promising that technological progress would bring about a more moral and peaceful world.

It was a seductive vision. One week after Reagan's speech, columnist Meg Greenfield felt she must "defer in some way, to the great weight of argument and opinion against" Reagan's proposal.[44] Yet she also challenged the moral authority of the argument:

Can anyone feel intellectually or morally content with a position that requires us all to assert, as a matter of national policy, that we are willing to obliterate millions upon millions of innocent, helpless human beings and cause others unimaginable suffering for any cause whatever?

It is an astonishment to me that 14 years after our own first landing on the Moon, and in an age habituated to mind-boggling scientific achievement . . . "Buck Rogers" and "Star Wars" should be dismissive terms of ridicule for a proposal such as Reagan's.

Historically, invention has succumbed to other invention, science has bested science. I wish the status quo nuclear gang would try to improve on Reagan's thought, not merely satirize it. I wish they, too, would think radically.[45]

Reagan soon invited scientists and engineers to think radically, commissioning a study of new defensive technologies. Led by former NASA director James Fletcher, the Defensive Technologies Study Team included over 60 scientists and engineers, and delivered a seven-volume report in the fall of 1983. Although the report concluded that a 99.9 percent effective defense was "not technically credible," panel members publicly expressed support for Reagan's ultimate goal.[46]

Reagan invited similarly radical thinking from strategists, with similar results. A study led by Fred Hoffman, an economist who had helped Wohlstetter analyze the vulnerability of bombers at RAND in the early 1950s, claimed that new technologies might eventually "provide a nearly leak-proof defense" and recommended focusing on limited defenses that would enhance deterrence, not eliminate it.[47] When the Fletcher and Hoffman studies were presented to Congress, the White House wrote an executive summary, promising: "The scientific community may indeed give the United States 'the means of rendering' the ballistic missile threat 'impotent and obsolete.'"[48]

1.4   Unprecedented Complexity and the Problem of Testing

Reagan used the Fletcher and Hoffman studies to justify a new Strategic Defense Initiative Organization (SDIO) in 1984. The SDIO absorbed BMD-related research from throughout the Defense Department, and reoriented much of it toward technology demonstrations that might hasten a deployment decision. Many scientists argued that the SDIO's flashy demonstrations, such as bouncing a laser beam off a mirror on the space shuttle, were a wasteful effort to gain credibility.[49] Such demonstrations were expensive, and the Reagan administration aimed to rapidly expand the SDIO's 1984 budget of $1.3 billion, to $2 billion in 1985 and upward, in a program that would cost a total of more than $24 billion by the end of the decade.[50] The enthusiastic SDIO director, General James Abrahamson, declared: "We can do in this country just about anything we set out to do technically . . . if we decide we want to do it."[51]

Such claims were soon challenged by a key question: how could the reliability of the system be proven? One of the Fletcher report's quiet conclusions was that the "problem of realistically testing an entire system, end-to-end, has no complete technical solution."[52] A panel on battle management computing and communications noted that short time-scales required highly automated decision making, and noted that there "is no technical way to design absolute safety, security, or survivability into the functions of weapons release and ordnance safety."[53] Instead, "Standards of adequacy must . . . be established by fiat, based upon an informed consensus and judgment of risks."[54]

While acknowledging challenges, the Fletcher panel did not concede defeat. Instead, it concluded that the "unprecedented complexity" of software required a new approach to weapons system development: "the battle management system must be designed as an integral part of the BMD system as a whole, not as an appliqué."[55] It recommended developing software with the managerial version of software engineering used widely in the Defense Department. The panel surveyed several Defense Department efforts to manage software more effectively, such as the Ada standard programming language, and the Software Technology for Adaptable Reliable Systems (STARS) program. Like the standard programming languages discussed in the previous chapter, STARS promised a "dramatic increase in software productivity" by reducing errors more quickly, yielding "more predictable software development times and costs."[56] The Fletcher panel proposed expanding these efforts by developing a new system for producing "warranteed software," including a "complete software development environment," "requirements and specification languages," "audit trails," and "knowledge-based support."[57] The panel suggested that with a good management system, software engineers could produce sufficient confidence in a space-based defense.

Other computer experts expressed skepticism. In the fall of 1984, two CPSR members, Greg Nelson and David Redell, warned that a "Star Wars" software system would be intrinsically unreliable. Citing disasters that had emerged only in operational use, and that had been documented in the pages of *Software Engineering Notes*, they rejected the notion that defensive systems could become credible by testing partial systems, saying "this technique is a totally inadequate substitute for testing under operational conditions."[58] Furthermore, they argued, "Since we have no spare planets on which to fight trial nuclear wars, operational testing of a global ABM system is impossible."[59]

As personal computers were growing more common, legislators also expressed skepticism. In the spring of 1985, the House of Representatives introduced legislation requiring Reagan to negotiate a ban on space weapons, and one of the bill's sponsors, Norman Dicks (D-WA), highlighted the software challenge. He contrasted a shuttle launch, which typically required 80,000 calculations in the final nine minutes before launch, with a comprehensive missile defense, which would require "over ten million calculations" in the same amount of time, noting that "it would have to work just right, the first time, without full scale testing."[60]

Meanwhile, arms controllers grew concerned that even limited tests of SDI technologies would encourage the Soviets to invest in offenses, making the ultimate challenge of defense even greater. In the days following Reagan's speech, Soviet leaders warned that Reagan's proposal would violate the ABM treaty's prohibition on field testing of "future" technologies such as lasers and space-based weapons. Though Reagan initially promised to abide by the ABM treaty, in April 1985, the SDIO announced its view that directed energy weapons and space-based components could be legally tested in the field.[61] Internally, Reagan's administration was in some disarray.[62] After conflicting public statements in the fall of 1985, it promised to adhere to what it called a "restrictive" treaty interpretation that prohibited field testing of lasers and space-based weapons, but also asserted that the still-classified negotiating record justified a broader interpretation of the ABM treaty.[63] Negotiators and legislators objected that such claims misconstrued the negotiating record, and neglected the congressional understanding that had been expressed during ratification.[64]

Reagan tried to persuade the Soviet Union to allow field testing of SDI technologies, insisting that defenses would not be threatening. He suggested that a future president might offer to share the defense, or even say, "I am willing to do away with all my missiles. You do away with all of yours."[65] Indeed, Reagan was so committed to SDI that when Gorbachev finally gave him the opportunity to eliminate nuclear missiles, he walked away. At a U.S.-Soviet summit in Reykjavik, Iceland, in October 1986, Gorbachev and Reagan discussed sweeping cuts to nuclear weapons, and even their elimination—a move that alarmed U.S. Defense Department and congressional officials who were committed to nuclear deterrence. But late in the day on October 12, talks broke down when Gorbachev insisted that Reagan affirm the previously shared interpretation of the ABM treaty. Reagan refused. The talks ended in anger.[66]

2 Physics, Computing, and the Limits of Progress

2.1 The Constraints of Nature

Jerome Wiesner was among many elite physicists who viewed Reagan's performance at Rejkavik as "confirmation, if any more were needed, that he is getting very poor advice."[67] In a *New York Times* op-ed piece, Wiesner and his fellow MIT physicist, Kosta Tsipis, blamed Star Wars on a small minority of physicists with the "Edward Teller" syndrome—the view that "any technical idea that surfaces" should be pursued.[68] Indeed, Reagan's science advisor, George Keyworth, was a Teller protégé. But with the Presidential Science Advisory Committee abolished, Wiesner and other former advisors had little or no direct access to the president. Instead, they opposed SDI through the public interest science groups and congressional advisory system that emerged in the wake of the 1960s missile defense debate.

For example, the Union of Concerned Scientists got to work almost immediately after Reagan's speech, sponsoring a panel on space-based missile defense that included luminaries such as Hans Bethe and Richard Garwin. Released just in time for the congressional hearings on SDI that opened in March 1984, their analysis assumed that proposed weapons would "perform as well as the constraints imposed by scientific law permit—that targets can be found instantly and aiming is perfect, that the battle management software is never in error, that all mirrors are optically perfect, that lasers with the required power output will become available, etc."[69] They showed that even in this "utopian" world, Soviet countermeasures could easily overwhelm new defensive technologies. [70]

A similar analysis emerged from the Congressional Office of Technology Assessment (OTA), chartered in response to the early 1970s controversies over the independence of the president's scientific advisers. Legislators wanted a second opinion on Reagan's vision of space-based defenses, and the OTA began its work by chartering a background study, to be completed by MIT physicist Ashton Carter. Like the UCS physicists, Carter assumed that efforts to engineer defensive technologies would be completely successful—that lasers would achieve the levels promised, and would perform perfectly in battle. Nonetheless, he used physical laws to demonstrate that countermeasures could likely overwhelm proposed space-based defenses, concluding that the prospect of a "leakproof" defense was "so remote that it should not serve as the basis of public expectation or national policy."[71]

The American Physical Society (APS) went beyond the UCS and OTA studies by analyzing the feasibility of the weapons themselves. In September 1983, the APS Panel on Public Affairs began working with Drell—who would soon become president of the APS—as well as Bethe, Garwin, and other elite physicists to form a study group on Directed Energy Weapons. After years of fundraising, classified briefings, and declassification reviews, APS finally released the report at its annual meeting in April 1987. The study group concluded that "even in the best of circumstances, a decade or more of intensive research would be required" to make any decision about deploying directed energy weapons.[72]

By the time the APS issued its study, a more formal OTA panel, including notable physicists such as Garwin and Drell, had completed a study analyzing the strategic implications of even limited defenses. In September 1985, the study warned: "Assured survival of the U.S. population appears impossible to achieve if the Soviets are determined to deny it to us."[73] Worse, a less than perfect space-based defense would likely create "severe instabilities."[74]

In fact, both the UCS and OTA studies showed that SDI technologies would likely cause two kinds of instabilities—those associated with an arms race, and those associated with crisis. Paul Nitze, Reagan's own advisor for arms control, acknowledged the potential problems. In a well-publicized 1985 speech, Nitze noted that very expensive defenses could give the Soviet Union an incentive to invest in relatively cheap offensive countermeasures, thereby creating arms race instabilities. He also acknowledged that if defenses could not survive a dedicated attack, they would become a tempting target for a preemptive first strike, especially in an uncertain and fast-moving crisis. Explaining what soon became known as the "Nitze criteria," he argued that defensive technologies should only be deployed if they were both "cost-effective at the margin" and "survivable."[75]

Though published analyses showed that SDI technologies would not likely be cost-effective or survivable, advocates of exotic defenses claimed that classified reports proved otherwise. Teller, Wood, and others working on SDI technologies accused the physicists associated with the UCS and OTA studies of "politicized science." Garwin and other physicists were quick to rebut their critics, but controversy grew.[76]

The rising debate became a training ground for a younger generation of physicists. Like the broader anti-nuclear movement, physicists with the Union of Concerned Scientists shifted from a 1970s opposition to nuclear power back toward nuclear arms control, and brought back the university "teach-ins" of the late 1960s.[77] At Cornell—home of UCS cofounder Kurt Gottfried as well as Hans Bethe and Carl Sagan—several physics graduate students and postdoctoral researchers became leaders in the renewed disarmament movement. In 1985, three of them—George Lewis, Lisbeth Gronlund, and David Wright—helped launch a nationwide boycott of SDI research. In a May 1986 press conference in Washington, D.C., they announced that 57 percent of the faculty in the top 20 physics departments in the United States had pledged not to conduct "Star Wars" research because a defense of populations was "not technically feasible in the foreseeable future," and SDI was "ill-conceived and dangerous."[78]

Although most of these younger physicists continued well-established traditions of analyzing defenses, by the 1980s a few devoted more attention to the challenges of battle management computing. In the fall of 1984,

Herbert Lin—a postdoctoral fellow at MIT's Center for International Security—began to analyze the computing challenge of space-based defenses.[79] Lin was painfully aware of programming difficulties from his computing-intensive graduate research, which he had recently completed with Manhattan Project veteran and arms control advocate Philip Morrison at MIT.

Lin had no formal training in computer science. But it was no longer the 1960s, when software difficulties were "well-remembered but not well documented," for software engineers had published recollections and data on large software projects. Lin turned first to Barry Boehm's *Software Engineering Economics*, a thick text that used data on more than 60 software projects at TRW to develop a Constructive Cost Model (COCOMO) of the time and effort required to develop various kinds of software.[80] Using the model for a large, highly constrained "embedded" project, Lin estimated that the SDI software would require between 13,400 and 81,700 "man-years" to develop.

Lin also grew familiar with the RISKS forum and CPSR, which highlighted the inadequacy of partial tests and small-scale testing and simulation for ensuring system reliability. In an article eventually published in *Scientific American*, Lin underscored the complexity of the software system for ballistic missile defense (figure 8.2). He noted that even "experienced software engineers" would inevitably discover and correct errors after such a complex system went into "operational use"—an impossibility for Star Wars.[81]

2.2   The "Essence" of Software: "Arbitrary Complexity"

Lin was still completing his study when Parnas resigned from an SDIO panel on battle management computing, explaining that the SDI software could never be trustworthy. Parnas's argument went beyond those based on engineering experience, by citing "fundamental" limits to the reliability of software. In the first of eight mini-papers, Parnas explained that computer systems could not be modeled by continuous mathematical functions that were familiar in engineering; instead they were represented by discrete logical functions. Furthermore, computer programs were so complex that they contained myriad paths through these logical functions. In principle, a computer could step through each path in a complex piece of software, and verify that it was logically correct. But in practice it would take the world's fastest computers millennia to completely verify even relatively simple programs. Hence, Parnas argued, software could only approach reliability as the most common paths through a program—those most often taken in operation—were tested and debugged through real-world use. In the next six papers, Parnas reviewed the requirements for SDI software, and explained why new tools, such as programming languages, programming environments, and

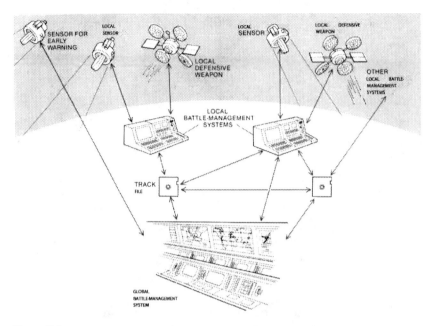

Figure 8.2
Battle management for a ballistic missile defense system would be exceedingly complex.
Source: Herbert Lin, "The Development of Software for Ballistic-Missile Defense," *Scientific American* 253, no. 6 (1985): 53.

formal program proofs, would not produce trustworthy software without operational testing. In short, he explained that "fundamental" limits on software reliability would "not disappear with improved technology."[82]

An eighth and final mini-paper critiqued the Defense Department's management of software writ large. Parnas recalled years of being "astounded at the amount of money" wasted on Defense Department software research, and "very large reports with very little content" that were "never used."[83] He blamed "technocrats" who had never been successful at software research and development, and thus were incapable of judging proposals.[84] Similarly, he claimed that each member of the panel from which he resigned lacked the experience needed to study the SDI software problem, yet each would "profit from continuation of the program."[85] He noted: "In the first sittings of our panel, I could see the dollar figures dazzling almost everyone."[86] Since SDI offered a "pot of gold" for pursuing "an unending set of technological puzzles that are fun to work on," Parnas warned: "You will find it very hard to find unbiased experts on this issue."[87]

Parnas recalled that when he sent his "octet" and resignation letter to the study directors, they tried to talk him into staying, promising that SDI funds would be good for research. Next, he sent his materials to the president's science advisor, Keyworth. This too, had little effect.[88] Parnas finally sent his letter to members of Congress. Soon, it was leaked to the *New York Times*.[89]

On July 12, 1985, the resignation letter began making headlines, sparking a firestorm of controversy in online forums like RISKS, as well as in publications such as *Software Engineering Notes* and the *Communications of the ACM*. Other members of the panel learned of Parnas's resignation though the newspapers, and began to grapple with the challenge to their credibility. They dubbed themselves the "Eastport Study Group," but remained better known as the "panel that Parnas resigned from," and grew determined to counter Parnas's public arguments. The computer science departments at MIT and Stanford each helped sponsor a debate. As CPSR members rallied around Parnas, their work gained more visibility than ever. And for the first time, Congress started including computing and software experts in its hearings on missile defense.

Parnas continued to assert that the panel was biased by the prospect of SDI research funding. In front of 1,300 people at the MIT debate, he recounted asking the Eastport Panel organizer, Dick Lau: "If we came to the conclusion that the thing really couldn't be done, could we say so?" To which he received the answer "NO," even though it was only lunchtime on the first day of meetings. Since Lau was in the audience, Parnas continued: "I challenge him to get up here and say he didn't say that to me." Lau protested loudly: "I said no such thing! . . . UNTRUE!!!" As the moderator tried to resume, Parnas insisted: "I remember the sandwich."[90]

Members of the Eastport panel were eventually investigated by the U.S. Congress for conflicts of interest, but more immediately, the panel confronted a challenge to its technical competence.[91] Parnas' critique struck home in at least one sense: none of the panelists was experienced in the problems of real-time military software that were essential to missile defense. The panel chairman, Danny Cohen of the University of Southern California, was involved with efforts to network computers. Charles Seitz, a computer science professor at Cal Tech and the nephew of the eminent physicist Frederick Seitz, was perhaps the most eloquent of the panelists, but he researched computer architectures. And though Princeton computer science professor Richard Lipton had earned his doctorate under Parnas, and coauthored the well-known critique of program proofs in the late 1970s, his work remained highly theoretical.

In an effort to address Parnas's critiques, the panel began to consult with software experts such as James Horning and Fred Brooks during the summer of 1985. Horning reported to members of the RISKS listserv that panelists weren't "crazies," "charlatans," or "fools," but recognized their lack of software expertise.[92] Horning was surprised by the panel's intense "belief that any technical problem could be steamrollered with a budget" as large as SDI's, and by its tendency to treat reliability as "one of the parameters that can be varied to make the problem easier."[93] Horning concluded that the feasibility of SDI software "all depends on how much functionality, coordination, and reliability are demanded."[94] But he was not optimistic: ". . . as with most other computer systems," he argued, "the dimension in which the major sacrifice will probably be made is reliability." He concluded that "with a system intended to be used 'at most once,' there may be no one around to care whether or not it functioned reliably."[95]

But what sense did it make to speak of a "reliable" system that had never been relied upon? This question simmered just beneath the surface of a debate that came to focus more ostensibly on notions of "fundamental mathematical" limits to reliability. It was a controversial claim for a community that remained committed to technological progress. Nonetheless, many computer experts reached similar conclusions.

Fred Brooks was leading a Defense Department software study when Parnas resigned, and as he reviewed decades of efforts in software engineering, he concluded that "no magical software technology" would yield a dramatic improvement in productivity or reliability.[96] In his 1970s book, *The Mythical Man-Month*, Brooks had warned that more "manpower" would not make complex software tractable. But by the 1980s, "more experience and more reflection yielded more insight" into what he termed the "essence" of software.[97] In what became a much-celebrated paper, Brooks explained:

> The physicist labors . . . in a firm faith that there are unifying principles to be found, whether in quarks or in unified field theories. Einstein argued that there must be simplified explanations of nature, because God is not capricious or arbitrary.
>
> No such faith comforts the software engineer. Much of the complexity that he must master is arbitrary complexity, forced without rhyme or reason by the many human institutions and systems to which his interfaces must conform.[98]

Brooks argued that "no silver bullet" would ever slay the werewolves of complex software because of its "irreducible essence"—arbitrary complexity.[99]

2.3   The Nature of Reliability: Mathematics and Murphy's Law

Brooks's argument, like Parnas's, sparked controversy in the computing community. Both were accused of pessimism, and both insisted that they were quite optimistic about progress in software engineering. In fact Brooks even expressed optimism about SDI software. In congressional testimony, he was not "so bold as to assert that the SDI software can surely be built," but believed "it can be conceivably done" because the SDI system did not need to be "absolutely perfect," merely "perfectly guarded against catastrophe."[100]

Parnas countered that the problem was not perfection, but reliability:

I drive a car every day that I trust and I know that it is not perfect. I just want it to be trustworthy. What I want to know is what its limits are, when it can succeed, and when it can fail. I want to have confidence that it will not fail catastrophically.[101]

Parnas explained that software systems "are never reliable when first put into use, their reliability increases after use, but you never really know that you have found the last bug."[102] He recalled stepping inside computer support trucks from the Vietnam War to find the walls covered with debugging notes. Unable to discover all the glitches before the system was actually used, programmers were sent into the field. Parnas emphasized: "We would not have that kind of opportunity with SDI."[103]

Members of the Eastport panel could not counter Parnas's claims of unreliable software by recounting their own experience developing complex software—they had little or none. Instead they disputed the general notion that fundamental laws made SDI software impossible. Cohen repeatedly insisted that "there is no fundamental principle that prevents meeting these [SDI software] requirements."[104]

Parnas answered that he had never claimed that it was impossible to develop SDI software—only that fundamental mathematics rendered it extremely improbable that the software would succeed in battle. He illustrated his point at a Stanford debate:

It's not impossible that this thing will be right the first time, it's not impossible that if you take 10,000 monkeys and let them type for 5 years that they'll recreate the Encyclopedia Britannica. . . .

Is there any fundamental reason why it's unlikely to work correctly when really needed? Yes, because we don't know what it's supposed to do, and we don't have any way to verify that it does what we think it should. . . . We have lots of experience in building software, we know what human beings can do and they can't build software that's right before it's really used. And we know good mathematical reasons why that's going to be the case.[105]

As this suggests, Parnas grounded his arguments in both engineering experience and fundamental mathematical analyses. Similarly, arguments about "trustworthiness" were inevitably social as well as technological, because they involved the judgment of users.

Such arguments were not sufficiently mathematical for many computer scientists. At the MIT debate, computer science professor and moderator

Michael Dertouzos suggested that the dispute would continue because computing possessed "no basic laws other than one, poor Turing's law," which offered little guidance.[106] By contrast, Parnas claimed that Shannon's mathematical theory of communications, which established mathematical limits to the ways that arbitrary bits of information could be compressed, was the "fundamental theory of computer science."[107] Such mathematical formalisms, he argued, proved his case: the software was too complex to be trustworthy without full system testing.

Ironically, members of the Eastport panel adopted a similarly mathematical approach when they concluded that trustworthy SDI software could be developed. Rather than addressing the social nature of reliability, they treated reliability as a probabilistic variable. In December 1985, the panel's final report argued that the SDI battle management software would be more likely to work properly if it were designed to operate in a "decentralized and loosely coordinated" manner.[108] By developing a system with many distributed elements that were only loosely coupled to one another, the software could be broken down into simpler, testable parts. The overall system would become more fault-tolerant and survivable, since errors or a direct attack on one element would not necessarily destroy the whole system.[109]

Panelists often used communications systems to illustrate the ways that loose coupling could improve reliability and survivability. For example, systems such as the ARPAnet were based on technologies that aimed at improving survivability in the case of a nuclear war. At a Stanford debate, Lipton suggested that communications systems were "pretty reliable" because they were "large collections of very loosely connected systems," or "metasystems."[110] But Parnas would admit no such explanation from his former student. He argued that in the mathematical definitions of systems theory, there is "no such thing as a metasystem that's not a system," and challenged Lipton to "make a mathematical definition of system that still makes sense, that makes a distinction between a system and a meta-system."[111]

Parnas soon drafted another paper: "Why Communications Systems Are Not Like SDI."[112] He emphasized that the telephone network, ARPAnet, and similar distributed systems had been developed only slowly, through continual use and testing. But for many, Parnas' critiques were persuasive less because of fundamental mathematics, than because of common sense. When Cohen emphasized that "no fundamental law" rendered software impossible at the MIT debate, a member of the audience objected: "You said building SDI does not contradict any fundamental law. . . . It does contradict Murphy's law."[113]

2.4   Simulation and the Risk of "Catastrophic Failure"

In the fall of 1985, the U.S. Congress instructed the Office of Technology Assessment to include software in its second study of space-based defenses. Whereas the first study panel was dominated by physicists, the second panel included more computer scientists, such as Parnas and Seitz. In 1988, the study group warned that because the system could not be fully tested, the "administration and Congress will have to decide on the deployment of a system whose performance would have to be predicted largely by computer simulations."[114]

By the time the OTA issued this conclusion, the SDIO's dependence on simulations had already become a practical problem. Weinberger announced an interest in near-term deployment in early 1987, but the specifics of the proposed deployment remained vague, because the Defense Department did not have simulation facilities that were powerful enough to even model the complexity of a future "Star Wars" type system. To enable such modeling, the SDIO had begun developing a new National Test Facility (NTF), an advanced computing center that would link a geographically distributed network of simulation facilities into a "National Test Bed," but this project soon stalled over contracting arrangements.[115] With Weinberger's interest in early deployment, the SDIO hastily moved computers into borrowed Air Force facilities near Colorado Springs in August 1987, but the SDIO did not expect the facility to become fully operational until late 1989 or early 1990.[116]

The SDIO's dependence on simulations also grew with controversy over the ABM treaty. SDIO director Abrahamson explained that with only "a set of tests here, a set of simulations here, a set of analysis here," the National Test Bed would become "a simulator of the overall system," enabling defensive research without violating the ABM treaty.[117] When Weinberger began to speak of early deployment in 1987, Senator Sam Nunn (D-GA), chair of the Armed Services Committee, warned that any effort to conduct tests outside of the so-called restrictive interpretation of the ABM treaty, without congressional consultation, would prompt deep cuts in the SDI budget. Furthermore, Nunn soon completed his own study of the treaty negotiating record, and accused the administration of a "complete and total misrepresentation."[118]

Even if the administration were allowed to breach the ABM treaty by field testing individual weapons, it would still be forced to rely on simulations to estimate the likely performance and cost of overall defensive system, and to choose the most promising system architecture. Without such estimates,

Weinberger's efforts at early deployment floundered in 1987. When SAGE veteran Robert Everett led a Defense Science Board (DSB) review of the SDIO deployment proposal, his task force deemed the plan too "sketchy" to be considered—"a list of components," but not "a consistent design."[119]

In 1988, Everett's DSB panel issued a more favorable assessment of the SDIO's revised deployment plan. But the panel again critiqued the tendency to think about missile defense as "a collection of major components . . . tied together by a Battle Management/C^3 system of some sort."[120] Everett explained in Senate testimony: "Instead of thinking of it as a collection of weapons," project managers should "think of it as a central sensor processing system into which you then add those reaction devices, weapons . . . as they make sense."[121] The *New York Times* explained that Everett's group emphasized "the brains, rather than the brawn, behind the Star Wars system."[122]

The DSB task force firmly supported the goal of a full, nationwide defense. It recommended beginning with a "limited deployment of long range ground based interceptors" that would provide "thin defense" against small attacks, perhaps focusing on key areas such as Washington, D.C.[123] Everett's panel also recognized that simulations could not supplant real-world testing, and concluded: "There is not a force acting on the SDI program that is more damaging or more insidious than the present debate on the 'narrow vs. broad' interpretation of the ABM treaty."[124] The panel felt that the debate had put the testing program in a "straitjacket," and recommended placing "the burden of proof on those who would restrain the program."[125]

Yet even if the program completely ignored the ABM treaty, only parts of a defensive system could be tested; the defense would inevitably rely upon simulations to build confidence in an overall system. And many software experts emphasized that any such confidence would be limited. Indeed, the most striking conclusion of the OTA study was summarized succinctly by the *Washington Post*: simulations and partial testing would have such limited use that the defense "likely would 'suffer a catastrophic failure' the first—and therefore only—time it was used."[126]

Both the OTA and the SDIO promptly objected to the *Post*'s synopsis.[127] The OTA report had not warned of a "likely" failure, but of a "significant probability (i.e., one large enough to take seriously)," of catastrophic failure.[128] Furthermore, the SDIO dissented from the OTA report, expressing optimism that "reliable, trustworthy software" could be developed. Nonetheless, the *Post*'s imprecision was telling. After decades of experience, the risks of failure in complex software had become a matter of common sense.

3 Conclusion: Software and the Rise of "Brilliant Pebbles"

By the end of Reagan's term, growing concerns about software were reshaping SDI. Confronted by growing scientific critique and dispute over the ABM treaty, the SDIO was forced to narrow its scope. The U.S. Congress cut more than thirty percent of the administration's request for both 1986 and 1987.[129] Sam Nunn had come to favor deployment of a limited system that would protect against an accident, but he also aimed to ensure that the SDI would comply with the DSB recommendations.[130]

In 1988, a new secretary of defense, Frank Carlucci, ordered SDIO Director Abrahamson to cut costs and to restructure the program in line with the recommendations of the Defense Science Board. Abrahamson initially proposed deploying space-based interceptors for as little as $40 billion, but by 1988 estimates rose to nearly $120 billion. Significantly, it was only by redesigning the defensive proposals around the challenge of computerized command and control that Abrahamson could cut budgets.[131]

Abrahamson's favored solution came from Livermore Labs, where physicists Lowell Wood, Gregory Canavan, and Edward Teller were looking for a successor to the x-ray laser—a weapon that was largely discredited by the end of 1988. Livermore's new scheme, "Brilliant Pebbles," would deploy thousands of missile interceptors, or Exoatmospheric Kill Vehicles (EKVs), in space. Wood emphasized that the miniaturization of microelectronics would make each EKV "smart": "Each pebble carries so much prior knowledge and detailed battle strategy and tactics, computes so swiftly and sees so well that it can perform its purely defensive mission adequately with no external supervision or coaching."[132]

In principle, Brilliant Pebbles would simplify battle management software, because it represented the ultimate loosely coupled system: thousands of kinetic energy weapons would be completely self-directed, requiring little or no coordination. When Abrahamson presented the new program to Congress in October 1988, he explained that rapid progress in computers had radically reduced the total cost of space-based interceptors, and had cut the cost of the battle management system in half. With "smart" EKVs and a simpler battle management system, Abrahamson estimated cutting the total program costs to around $70 billion.[133]

When President George H. W. Bush took office in 1989, his administration made Brilliant Pebbles the focus of SDI. Bush's administration finally acknowledged what many scientists and engineers had declared for years: a leakproof defense was not achievable. Secretary of Defense John Tower proposed cuts in Reagan's recommended SDI budget, and stated plainly:

". . . I don't believe that we can devise an umbrella that can protect the entire American population from nuclear incineration."[134] Nonetheless, the administration continued to press for a rapid deployment of Brilliant Pebbles.

As we will see in the next chapter, shrinking microprocessors failed to make Brilliant Pebbles "smart." Instead, EKVs came to depend upon complex networks of surveillance and tracking satellites for instructions—and these would remain mired in software challenges for decades. Meanwhile, as disputes about the ABM treaty continued, efforts to accelerate deployment required reliance on increasingly complex defense simulations. Scientists never would agree upon the feasibility of SDI software. But by the late 1980s, computer experts did agree that the information system must come first, before exotic weaponry. The computer technologies and systems recommended by computer experts would outlive Reagan's dream, and help propel missile defenses into a new era.

9   Conclusion: Complexity Unbound

Our new approach [to missile defense] will . . . deploy technologies that are proven and cost-effective and that counter the current threat, and do so sooner than the previous program. Because our approach will be phased and adaptive, we will retain the flexibility to adjust and enhance our defenses as the threat and technology continue to evolve.[1]

—President Barack Obama, September 17, 2009

In some ways, the Star Wars era of missile defense seems distant today. The 1980s sketches of how lasers might someday destroy Soviet missiles in space were replaced years ago, by real-life videos of Patriot interceptors lighting up skies in the Middle East, seeming to destroy Iraqi missiles perfectly. Today, U.S. defenses aim to intercept a very few missiles that might be launched from North Korea or Iran, not thousands of Soviet missiles. To many Americans, defenses seem like good insurance against a distant, but inevitable threat.

Yet, in other ways, the promise and risks of the Star Wars era are closer than ever. Despite decades of arms reductions, Russia and the United States each still deploy thousands of launch-ready nuclear warheads, maintain thousands more in stockpiles, and support armies of nuclear weapons scientists who design newer variants.[2] We carry great risks with these forces— of accidents, unauthorized attacks, and the continued spread of nuclear technologies to terrorists and other nations. Meanwhile, efforts to negotiate reductions continue to hang on defenses—for while Russia may agree to reduce offenses bilaterally, it will not allow U.S. defenses to undermine its prowess unilaterally.

What should we make of this strange, new-old world? After more than six decades of debate about missile defense—whether it will facilitate arms control or encourage offensive development, insure against attacks or spark a conflict, prove reliable or fail catastrophically—what have we learned about the promises and risks of defense? And most important, how can defenses be best designed and deployed to reduce risks?

This chapter shows how decades of argument about missile defense have culminated in efforts to design, adapt, and test complex defensive systems since the Star Wars era. It builds on earlier chapters of this book, using an excellent set of recent analyses by the most visible scientific group in today's debate: physicists. A growing number of physical analyses suggest that limited defenses can in principle be quite useful. To be sure, many physicists argue that U.S. defenses have not been tested sufficiently for deployment, and virtually all agree that any defense can be overwhelmed with sufficiently complex countermeasures. But they no longer assume that defenses will trigger an arms race, or create instabilities in times of crisis. In fact, some of the most prominent physicists we have discussed, such as Sidney Drell, have returned to the logic of the early Cold War, arguing that today cooperative defenses could assure the United States and Russia that it is safe to cut numbers of nuclear warheads drastically—perhaps even to zero.[3]

Thus, when we analyze defenses as physical systems, they have brighter prospects today than they did during the Cold War. But when we consider defenses as information systems, a very different picture emerges. Today's defenses confront fewer physical missiles, but greater informational complexity—a wider range of threats, emerging from a larger number of regions, in uncertain and changing ways. Defense in this multi-polar world also requires more international cooperation than ever, as the United States must station radars, missiles, and other assets in regions around the world. Cooperation is not merely a diplomatic nicety—it is essential to the technical effectiveness of today's defenses. These technical requirements constantly shift with emerging threats, domestic politics, and negotiations among nations whose interests sometimes align, but sometimes diverge.

Such complex and ever-evolving requirements will limit any engineer. But software engineers have come to recognize this type of complexity as their "essential" challenge. Indeed, the fundamental engineering challenge confronted by today's defenses may be most aptly captured by Fred Brooks's notion of "arbitrary complexity," discussed in the previous chapter. Brooks argues that the complexity of software is fundamentally different from that of physics. Whereas the complexity of the physical world is assumed to reflect a hidden logic, arbitrary complexity is "forced without rhyme or reason by the many human institutions and systems to which [software] interfaces must conform," and which perpetually change.[4]

As we will see, the arbitrary complexity associated with rapidly deploying defenses to new and changing environments has created new risks since the Star Wars era. Furthermore, the challenges of arbitrary complexity make it impossible to have defenses that are both "proven" and "adaptive." Defenses that seemed reliable in one context have repeatedly failed when rapidly adapted to a new and unpredictable environment. To illustrate the risks associated with efforts to rapidly deploy defenses, we first turn to the fading years of the Star Wars program, when a declining Soviet Union turned missile defense politics upside down.

1   Defense in the Era of Video-Game War

1.1   "Triumph of Silicon over Steel"?

Missile defense politics began to shift dramatically in 1989, when revolutions in six different countries, all of them previously operating as Soviet satellites, suggested that the Soviet empire was on the decline. The world watched Berliners ripping down the wall dividing East from West on November 9, 1989, dancing and pouring through breaches.[5] Less than a month later, at a summit meeting in Malta, U.S. president George H. W. Bush called for an end to the Cold War, and Soviet leader Mikhail Gorbachev declared that a new era had begun. Bush remained cautious, but to many observers it was obvious: the Cold War was over.[6]

In the United States, many pundits and politicians began to argue that Reagan had cleverly bankrupted the Soviet Union with SDI, forcing the Soviets into an expensive competition in defenses.[7] As Pavel Podvig, Peter Westwick, and others have discussed, the historical record provides little support for this view.[8] But SDI supporters needed a success story, especially as the Soviet decline left them with a new identity crisis: what threat would SDI counter?

They soon found an answer in threats from "rogue" states such as Iraq or North Korea.[9] And when Iraq invaded Kuwait in August 1990, the new mission was put to the test. In January 1991, Iraq retaliated against the bombing of a U.S.-led coalition, by launching Scud intermediate-range missiles toward Israel. The United States responded by rushing the Army's Patriot defensive systems to Saudi Arabia and Israel. The Patriot battalions— including radars, interceptors, and computerized control systems manned by crews in a truck—were originally designed to intercept aircraft flying up to a speed of Mach 2, but engineers quickly reprogrammed them to target Scuds at a speed closer to Mach 6. Throughout January 1990, television footage showed Scud missiles and Patriot interceptors streaking toward one another through the night sky, colliding in a brilliant fireball, and disintegrating far above the ground. Defenses seemed to have readily adapted to, and triumphed in, the new mission.[10]

For many, the Patriot's performance was just one example of an information-driven revolution in warfare. The *New York Times* declared "Hero Status for the Computer Chip," and news broadcasts showcased several highly computerized systems, including the Patriot, laser-guided bombs and Stealth fighters.[11] At the Federation of American Scientists, John Pike declared "the triumph of silicon over steel," while Sam Nunn declared "a new era of warfare."[12] Industry engineers claimed that the United States could now direct a "devastating air attack into the heart of populated areas without the risk of causing civilian casualties."[13]

Other observers criticized "video-game" war for turning violence into spectacle.[14] And high-tech had its limits. Iraqi leader Saddam Hussein accused Western forces of hiding behind high technology, calling it American cowardice.[15] Because Hussein located troops near schools and archeological treasures, the Air Force was unable to avoid killing civilians. After Hussein failed to accept United Nations resolutions calling on him to withdraw, the coalition finally sent in ground troops, and it was the ground campaign that drove Hussein to accept the resolutions in just two days.[16]

Furthermore, it soon emerged that computer systems had not performed as well as first advertised during the Gulf War. Initially, missile defense advocates argued that the widely-broadcast Patriot interceptions proved that "SDI's software nay-sayers were wrong."[17] As debate heated up in the RISKS forum, David Parnas emphasized that space-based defenses faced much more complex challenges than the Patriot.[18] And as it soon emerged, the Patriot could not handle even the more limited threat of Scud attacks. Software problems became evident in late January, when the Patriot fired two interceptors at U.S. fighter jets.[19] The fighters evaded the attack, but other defects carried more tragic consequences. On February 25, 1991, a software-controlled timing error caused the Patriot to miss a Scud missile, which struck an Army barracks at Dhahran, Saudi Arabia, killing 28 Americans. The necessary software fix arrived in Dhahran the next day.[20]

The failure underscored Parnas' central argument in the SDI debate: software only approaches reliability through operational testing and debugging, and the unexpected can always emerge after years of use. SDI director Henry Cooper later acknowledged that the Patriot was rushed into war such that "people witnessed operational tests and evaluations on CNN."[21] Indeed, the software was modified six times during the 1990 war.[22]

In fact, physicists soon showed that the unpredictable operating conditions contributed to a much more massive failure of the Patriot. George Lewis, who had helped lead a nationwide boycott of Star Wars research as a physics grad student at Cornell, joined forces with Theodore Postol, a physicist who had served as advisor to the chief of naval operations, to analyze television footage of the Patriot "interceptions."[23] They showed that most of the interceptors missed the warhead—in part because the Scuds reentered the atmosphere at a higher speed than anticipated, and frequently broke up upon reentry, spiraling downward rather than following a predictable ballistic trajectory.[24]

The Army continued to insist that the Patriot had been successful. And although the American Physical Society eventually confirmed the accuracy of Postol's and Lewis's critique, illusions of the Patriot's success persisted, buoying those who supported an early deployment of SDI technologies.[25] Robert Cooper, the SDIO director, pointed to the Patriot's defense of Tel Aviv: "I don't mean to be flip in this, but I believe that the people of New York City and Indianapolis and Honolulu, Hawaii, deserve that protection, too."[26] Bush highlighted the "remarkable" success of Patriot in his State of the Union Address at the end of 1991, announcing that the SDI mission would shift toward "Global Protection Against Limited Strikes" (GPALS). In the new era, SDI would no longer aim to counter large numbers of missiles coming from the Soviet Union, and would instead attempt to counter a wider range of ballistic missiles, launched from any nation, toward the United States or its allies.[27]

1.2   Missile Defense Politics and the Evolving Software Problem

Though Bush's rhetoric linked the Patriot to SDI, the military services had been developing such theater defenses—which would protect deployed forces from short and intermediate range missiles—for decades. Theater defenses had never received support under Reagan's Star Wars program, and they did not become a focus of Bush's redirected SDI. Instead, Bush proposed deploying Exo-atmospheric Kill Vehicles (EKVs), designed to collide with, and thereby "kill" missiles during their midcourse of flight. He urged the U.S. Congress to authorize an initial deployment of Ground Based Interceptors that would launch EKVs into space. He also requested the option to later deploy tens or hundreds of thousands of the space-based "smart" EKVs, thereby continuing the Brilliant Pebbles scheme discussed in the previous chapter.[28] As congressional interest in theater defenses grew, Cooper began claiming that Brilliant Pebbles would be able to intercept any missile with a range of more than 300 miles, including Scuds.[29]

The continued focus on space-based systems complicated the software for both theater defenses and National Missile Defenses (NMD). In principle, Brilliant Pebbles would make the defensive software simpler, by using large numbers of smart, self-directed EKVs that could function without complex coordination. But in practice, the challenges of basing EKVs in space encouraged the development of rather weak "eyes"— mercury-cadmium-telluride, or "mercad" sensors—and thus put more demands on the software "brains" of Brilliant Pebbles.

The Army had begun developing a much more capable silicon sensor as part of its ground-based midcourse experiments in the early 1980s. The silicon sensor "could see a normal ice cube out of a refrigerator at 75 miles," and could detect a wide range of infrared wavelengths, a key advantage in discriminating warheads from decoys.[30] Thus, once ground-based radars guided the EKV near the target, its silicon "eyes" would need relatively little additional help to find the warhead among decoys and home in for a kill. However, Mercad was preferred for space-based defenses because it only required cooling to 70 Kelvin, whereas the silicon sensor required cooling to 10 Kelvin—something that could only be done from the ground before launch. Unfortunately, the relatively myopic mercad EKVs would require multiple sets of directions—first from ground-based radars, and then from a complex network of surveillance and tracking satellites—to get within striking range of its target. Integrating data gathered from these different orientations seriously complicated the missile-tracking software.[31]

Nonetheless, advocates of space-based defenses ignored these software challenges in the early 1990s. It was the least of their worries, as they met stiff opposition and a tumultuous political environment. Richard Garwin was among physicists who argued that even if Brilliant Pebbles were viable, theater defenses would be much more effective from the ground than from space.[32] And many members of Congress argued that theater defenses were the far more urgent priority. Furthermore, they argued that space-based defenses were a violation of the ABM treaty, a threat to satellites, and a sticking point in negotiations to reduce nuclear arsenals. Although Gorbachev and Bush signed the Strategic Arms Reduction Treaty on July 31, 1991, agreeing to reduce offensive nuclear weapons by more than 20 percent, reductions remained tenuous. The dissolution of the Soviet Union and rise of Boris Yeltsin as the first president of the new Russia in 1991 delayed ratification of the treaty for over a year.[33] Eager to reap a "peace dividend" by ending the arms race, Democratic majorities in Congress deferred any commitment to space-based interceptors and cut funding for Brilliant Pebbles.[34]

Despite opposition to space-based defenses, support for National Missile Defenses (NMD) grew after the Patriot's apparent success. The Missile Defense Act of 1991 included an ambiguous mandate to deploy NMD "by the earliest date allowed by the availability of appropriate technology."[35]

When President Bill Clinton inherited this mandate in 1993, he brought NMD down to earth. He more than doubled proposed funding for the ground-based midcourse interceptor. By contrast, he cut Brilliant Pebbles funding to just one-third of previous levels and relegated it to long-term research, while affirming the narrow interpretation of the ABM treaty.[36]

Clinton also made theater missile defenses the top priority of the Ballistic Missile Defense Organization (BMDO), his redirected SDIO.[37] For example, the Navy's ship-based Aegis missile defense system fared well in the new BMDO's developmental programs, along with the Army's Patriot and Theater High Altitude Area Defense (THAAD) systems.[38] The Air Force's airborne laser program, which aimed to integrate high-energy lasers into an airplane so they could zap missiles in their boost phase, found a place in long term research.[39] As we will see, many of the theater defense programs supported under Clinton were eventually advocated under the rubric of "flexible" and "adaptive" approaches to NMD.

1.3 "Dry-Labbed" Data

Throughout the 1990s, the growing emphasis on regional conflicts, combined with expanded funding for theater defenses, increased demands for defensive software that could be rapidly deployed in operations around the world. By contrast, the National Missile Defense program only tested experimental defenses. Nonetheless as contractors began testing software for midcourse interception, they found a much greater challenge than first imagined.

In 1990, the Defense Department awarded three different companies contracts to competitively research and develop a ground-based interceptor and its Exoatmospheric Kill Vehicle (EKV). Rockwell, one of two companies to survive the initial competition, developed an EKV design based upon the silicon sensor. The other company, Hughes, opted for Mercad. Both approaches were soon beset with problems, and in 1993, Clinton slowed efforts to issue a production contract, demanding more research, development, and testing.[40]

The EKV based on the silicon sensor had the simplest software job. It would require only information from ground-based radars to approach the target. In principle, the high-capability silicon sensor would then allow the EKV to home in on the correct target. Boeing, which eventually took over the project from Rockwell, subcontracted the EKV software to TRW. But in February 1996, Nira Schwartz, a TRW engineer testing the signal processing software, concluded that the silicon-based EKV could not successfully discriminate real warheads from decoys. Schwartz insisted that TRW report the problem both to the Defense Department and to Rockwell, the prime contractor for the ground-based interceptor. The next month, she was fired. Schwartz filed a suit under the False Claims Act of 1996, prompting the U.S. Justice Department to open an investigation.[41]

While the investigation wore on, the BMDO conducted two Integrated Flight Tests in an effort to select a contractor for the Ground Based Interceptor. In June 1997, TRW's EKV flew by a suite of targets in space in an effort to discriminate a warhead from a suite of other targets. Both TRW and the BMDO claimed that the EKV and its algorithms successfully identified the warhead. Schwartz objected. She had strong support from another engineer who had also worked on the EKV software, and who testified that TRW's analysis was "fudged, dry-labbed, manipulated and censored" to get the desired result.[42]

Despite such testimony, the Justice Department declined to join Schwartz's case in March 1999. Thus, Schwartz went to journalist William Broad, who soon presented her case on the front page of the *New York Times*.[43] Publicity only grew after Postol read about Schwartz in the *Times*, and took up her cause.[44] Schwartz also contacted the congressman who had authored the False Claims Act, Howard Berman (D-CA), who requested an investigation by the General Accounting Office. Although the GAO did not find TRW guilty of fraud, it criticized the company for failing to fully document the limitations of the integrated flight test. Instead TRW characterized the sensor's performance as excellent and called the test a "success," terms that could easily be "misunderstood."[45]

By the time the software controversy made headlines, TRW was out of the running for the EKV contract. Members of TRW's team had been discovered with documents from the rival team, whose project had moved from Hughes to Raytheon. Boeing attempted to take corrective action, but nothing would suffice. The competition was canceled in December 1998, leaving Raytheon as the default contractor.[46]

TRW's elimination only increased the software challenges for midcourse defense. Raytheon's EKV used the relatively "dumb" Mercad sensor that would require guidance from satellite surveillance and tracking systems. And by the late 1990s, decades of efforts to upgrade the Defense Department's 1970s-era satellite system were still behind schedule, over cost, and mired in software challenges.[47]

2   A Rush to Failure

2.1   Accelerated Testing and the Rise of Simulation

The BMDO eventually came under fire for not reopening the competition in search of a better EKV design.[48] But a new competition would take time, and by the late 1990s, program managers felt intense political pressure to get on with deployment. Clinton's slow-go approach to deployment lost many congressional supporters in 1995, when Republicans won majorities in both the House and the Senate. That year, Congress sent Clinton a Defense Authorization Bill mandating an NMD deployment by 2003. Clinton vetoed the bill, but pressures only grew.[49]

Many defense conservatives rejected the 1995 National Intelligence Estimate's conclusion that no new countries would "develop or otherwise acquire a ballistic missile in the next 15 years that could threaten the contiguous 48 states and Canada."[50] Congressional critics commissioned a reassessment led by CIA director Robert Gates, but in December 1996, Gates's panel concluded that evidence against a rogue state acquiring ICBMs had only grown stronger. Congressional conservatives insistently commissioned another study, this one chaired by Donald Rumsfeld, a long-standing advisor of Reagan and other Republicans and vocal supporter of missile defense. In July 1998, Rumsfeld's report concluded that within five years of deciding to develop an ICBM, a rogue state would be able to "inflict major destruction" on the United States.[51] The report's timing was perfect. Four months later, North Korea launched a medium-range Taepo-dong missile. Though the missile's third booster phase failed to ignite, intelligence officials concluded that North Korea was closer to building an ICBM than previously believed.[52]

By that time, Clinton had begun responding to political pressures by upgrading the National Missile Defense from a "technology readiness" to a "deployment readiness" program. Under the "3+3" plan, launched in April 1996, Clinton aimed to make a deployment decision in three years, allowing some form of a national missile defense to achieve initial operating capability in another three years, by 2003.[53]

But just one year after the plan was launched, its estimated cost had doubled from $2.3 billion to $4.6 billion, and it was behind schedule.[54] Congressional investigators later explained that the program "was not sufficiently defined for detailed cost estimating" when it was moved from a technology readiness program to a deployment readiness program.[55] Furthermore, BMDO director general Lester Lyles worried that tight budgets and schedules were reducing "the rigor, robustness and number of tests" that should attend "a fully built-up program."[56] Indeed, the rush forced managers to plan 20 tests in 24 months, leaving little room to learn from failures.[57]

Grappling with shortened schedules, Lyles explained that the BMDO would mitigate risk through modeling and simulation, "to make sure we're not completely relying on a potentially random [successful] intercept."[58] Indeed, ground tests, simulations, and analysis became crucial to accelerating deployment of both Theater and National Missile Defenses in the early 1990s. The services maintained distinctive simulation facilities, but efforts to rapidly field theater defenses spurred the National Test Facility, begun under Star Wars, to begin modeling the interoperability of diverse systems like THAAD, Aegis, and Patriot.[59] The BMDO also used the facility for war gaming exercises, helping test proposed battle management and control systems for National Missile Defenses.[60]

In the 1990s, the Army also began developing simulation facilities at its Advanced Technology Center in Huntsville, Alabama, which would use data from components of a National Missile Defense to conduct Integrated Ground Tests (IGTs) of how the entire system might function. The Defense Department's Director of Operational Testing and Engineering, Philip Coyle, explained that Integrated Ground Tests were crucial to projecting the likely performance of defense against a wide range of threats, because the "nature of strategic ballistic missile defense" made it "impractical to conduct operationally realistic intercept flight testing across the wide spectrum of possible scenarios."[61] Yet by the late 1990s, the simulation capabilities were still "too immature to provide reliable estimates of performance."[62]

Furthermore, the NMD battle management software remained problematic. When Coyle participated in a 1999 battle planning exercise at the National Test Facility, he observed that the software did not provide operators with adequate situational awareness. Nor did it integrate the defensive radars and weapons properly. For example, it replicated rather than merged each target that was tracked by multiple radars, thereby producing "phantom tracks" that were automatically assigned to weapons.[63]

Coyle echoed the president of the Institute for Defense Analysis, retired Air Force general Larry Welch, whom he'd asked to lead a study of efforts to field missile defenses. In March 1998, Welch recommended slowing and restructuring NMD to allow more time for simulations and ground testing. When Welch testified before the U.S. Congress, advocates of defense asked if more funding wouldn't "accelerate this thing a little bit quicker?"[64] Welch objected that this approach forced "a test program that was simply inexecutable," a "'program managers' wheel of torture."[65] In a warning that would resound for years Welch argued that accelerating deployment would produce a "rush to failure."[66]

Secretary of Defense William Cohen responded by slipping the deployment readiness review by one year, and the earliest deployment date from 2003 to 2005.[67] But this hardly slowed the rush. Preparing for a June 2000 readiness review, the BMDO planned three flight tests of the EKV in just eight months.[68]

On October 4, 1999, the NMD program featured Raytheon's EKV in its third Integrated Flight Test, the first effort to actually intercept a warhead. Program managers were elated when the EKV intercepted a mock warhead, seeming to distinguish it from a balloon decoy. But critics later noted that the balloon was an obvious fake whose large size and bright signals had actually helped guide the EKV toward the nearby warhead.[69] And in January 2000, when a fourth Integrated Flight Test incorporated more system components, increasing the complexity of the test, the EKV failed. The third test suddenly seemed decisive. As publicity grew, program managers postponed the next test and added an extra communication system to ensure success. Nonetheless, in July 2000, the test failed when the EKV did not separate from the rocket.[70]

2.2   A Clean Break or Cold War Renewed?

When the Clinton administration finally reviewed the NMD program in August 2000, it was concerned about far more than failed tests. Russian leaders had long warned that the deployment would violate the ABM treaty and threaten negotiations to reduce nuclear arsenals.[71] Nonetheless, on July 23, 1999, Clinton signed a new National Missile Defense Act that committed the United States to deploying NMD "as soon as technologically possible," dropping his earlier insistence that a deployment decision include diplomatic considerations.[72] The stage was set for a showdown.

In August 1999, Russian president Boris Yeltsin suggested that the treaty was open to discussion. Clinton offered to help Russia finish building an early warning radar in exchange for an amendment that might permit an NMD deployment.[73] But two months later, Russia rejected the offer.[74] Furthermore, Yeltsin's successor as president, Vladimir Putin, soon warned that if the U.S. were to deploy its NMD, Russia would withdraw from existing treaties and stop cooperating on non-proliferation.[75] Putin stated that a U.S. NMD would "signify the undermining of the global balance," a statement that received strong backing from China.[76] On September 1, 2000, with nuclear treaties threatened and a lousy testing record, Clinton announced that the deployment decision would be deferred and passed on to a new president.[77]

Just days after Clinton's announcement, Coyle continued to urge that NMD development be slowed.[78] But President George W. Bush and his secretary of defense, Donald Rumsfeld, threw caution to the wind, announcing plans to "deploy defenses as soon as possible" in the spring of 2001.[79] In a widely publicized speech at the National Defense University, Bush presented a sweeping vision: layered defenses would protect the United States, its allies, and forces abroad from every kind of missile, while a newly cooperative relationship with Russia would combine limited defenses with mutual cuts in offenses. Where Clinton had sought to renegotiate the ABM treaty, Bush sought a "clean break" with the past.[80]

Bush was soon traveling around the world to enlist the support of Russian, Canadian, and European leaders. None rejected Bush's vision outright. The administration did not propose any specific plan, so nobody knew just what was on the table. But all nations cautioned the Bush administration against unilateral action.[81] International cooperation grew in the aftermath of the terrorist attacks of September 11, 2001, and world leaders remained open to dialogue.

But it soon became clear that Bush was less interested in dialogue than in persuading others to follow along. After hopeful negotiations in October 2001, Russia and the United States reached an impasse at a meeting in Moscow on November 3, 2001. There, Rumsfeld outlined ambitious plans to test missile defenses on land, sea, air, and space. Administration officials later recounted Putin's objection: "You want such flexibility that you are asking to effectively gut the treaty."[82] John Bolton, undersecretary of state for arms control, who joined the meeting, brashly agreed—the testing program would be unrestrained by the treaty. Putin called the demand unacceptable. "Well then," Rumsfeld reportedly replied, "we will be giving you notification of unilateral withdrawal from the treaty."[83] Additional conversations saw no agreement. On December 12, 2001, Bush announced that the United States would withdraw from the ABM treaty.[84]

Bush's announcement gave the U.S. legal freedom from all treaty constraints after a six-month waiting period. One day after the United States formally abandoned the treaty in June 2002, Russia announced that it would no longer abide by the Strategic Arms Limitations Agreements (START II), under which both superpowers were to de-MIRV their offensive missiles.[85]

Meanwhile, Bush asked Congress for $7.6 billion so that the new Missile Defense Agency (MDA), Bush's version of the BMDO, could begin deploying National Missile Defenses. The request met some opposition. It was still unclear what technologies or system would be deployed, and many leaders argued that the money would be better spent defending the United States from suitcase nukes and other terrorist threats.[86] But it was difficult to oppose any defense program in the wake of September 11, and the Bush administration won its request. In December 2002, Bush announced initial plans to deploy six ground-based interceptors (GBIs) at Fort Greely, Alaska, and four GBIs at Vandenberg, California, by 2004. He explained that more interceptors would follow, and the midcourse defenses might eventually be supplemented by defenses to intercept missiles in their boost and terminal phases.[87]

National Missile Defenses wrestled with far more than the integration of U.S. forces—they required international cooperation. To track North Korean ICBMs in their midcourse phase, as they traveled over the North Pole, the midcourse defense would rely upon far northern radars. The administration succeeded in gaining British approval to upgrade its early warning radar at the Royal Air Force Base of Fylingdales, and Danish approval for upgrades to a radar at Thule, Greenland.[88] But when the administration sought to counter an Iranian threat with interceptors in Poland, and a radar network in the Czech Republic, it met stiff resistance. Though the United States insisted that the system would target small threats from Iran, it would undoubtedly have some capability against Russian ICBMs. Russian leaders repeatedly expressed anger about U.S. plans, and pressured Poland and the Czech Republic to refuse cooperation. Thus, Eastern European countries balked at the U.S. request, seeking some guarantees of security from the Russian giant.[89]

Despite ongoing U.S.-Russian negotiations, tensions grew and erupted into something approaching a proxy war in Georgia. The United States had enlisted Georgia as an ally during the 2003 Iraq invasion, and persistently lobbied for its inclusion in NATO. These actions built up Georgia's military strength and inadvertently encouraged Georgian assertiveness against South Ossetia—a nation strongly backed by Russia. In early April 2008, NATO refused U.S. pressure to admit Georgia, but agreed to support the U.S. missile defense deployments in Europe.[90] Russian anger flared. In one of his last acts before passing the Russian presidency to his close ally, Dimitri Medvedev, Putin soon pledged increased support for South Ossetia. Skirmishes near the border of Georgia and South Ossetia escalated in the summer of 2008, and on August 7, Georgian troops launched an offensive into South Ossetia. The next day, Russia launched airstrikes on Georgia and began rolling tanks into South Ossetia.[91]

Within a week, Russia and Georgia agreed to withdraw troops to areas occupied before the outbreak of fighting.[92] Nonetheless, Russia continued sending forces into Georgia, seeming to expect expanded power in the region.[93] When the United States demanded an immediate pullout and sent humanitarian aid, Russia defiantly tightened its hold.[94] Poland, growing anxious about Russian territorial ambitions, finally agreed to allow the United States to deploy its Ground Based Interceptors on its soil in exchange for protection by Patriot missile defense battalions. Russian leaders responded with vows to break off all military cooperation with NATO.[95] Seven years after Bush announced a "clean break" with the past, it seemed that the Cold War had only renewed.

2.3 Testing and the Hurried Evolution of Defense

The Bush administration maintained that missile defenses were worth the diplomatic cost. Indeed, the administration argued that the missile threat was so urgent that National Missile Defenses should be deployed before completing the Initial Operational Test and Evaluation (IOT&E) process, wherein the entire system, including production-line components and trained operators, would undergo especially "realistic" testing. Though U.S. defense acquisition laws required that any major weapons system pass IOT&E before deployment, Bush planned an exception in order to deploy defenses rapidly. As we have seen, Clinton's director of OT&E, Philip Coyle, had already urged a slower testing pace. And in March 2004, Bush's director of OT&E, Thomas Christie, acknowledged that the missile defense agency would not conduct such realistic tests "for the foreseeable future."[96] Nonetheless, the administration aimed to begin operational deployment just six months later.

The Bush administration argued that testing and deployment could be combined under an evolutionary acquisition approach, in which there would be no "final, fixed missile defense architecture," but rather an "initial set of capabilities that will evolve to meet the changing threat."[97] Yet weapons procurement laws actually mandated an initial operational testing phase for each increment in an evolutionary acquisition process.[98] Furthermore, in the fall of 2002, the Defense Science Board advised the administration that it must define some system architecture to even get started.[99] Bush took a step in this direction by announcing that an initial ground-based midcourse system would form the basis of the NMD "Block 2004" capability.[100] Additional capabilities would be added in two-year increments, but no initial operational tests were scheduled, only highly constrained field tests.

Physicists repeatedly demonstrated the limits of these tests. Well before Bush decided to deploy the midcourse defense, two physicists at the Union of Concerned Scientists (UCS)—Lisbeth Gronlund and David Wright— orchestrated a study of the NMD viability against countermeasures. Gronlund and Wright had worked with George Lewis on the Star Wars research boycott in the 1980s, and became members of the UCS research staff in the 1990s. Working with physicists such as Lewis, Kurt Gottfried, Richard Garwin, and Ted Postol, they conducted a comprehensive analysis of countermeasures. In May 2000, their final report noted that the Defense Department's midcourse interception tests only succeeded when they used relatively predictable and simple targets.[101] Four years later, they updated their analysis, showing that tests of the EKV interceptor used warheads and decoys with remarkably different physical signals, making it easy for the EKV to find the "real" target.[102] Furthermore, the system often failed even these easy tests. Indeed, by the end of 2004, only five of nine intercept attempts could be construed as successes.[103]

Most physicists' studies emphasized that the midcourse defenses were limited by the laws of nature. But many tests failed even before confronting physical limitations—they struggled against the limits of complexity, and especially software.[104] For example, in December 2004, the first full system test of the NMD failed when the Ground Based Interceptor didn't launch. The director of the Missile Defense Agency, Lieutenant General Henry Obering III, dismissed the failure as the result of a "very minor" software glitch that could be easily corrected.[105] But how many glitches had yet to be discovered? Obering didn't raise the question, but insisted that the initial system "would work" against a simple attack.[106]

Congressional investigators were less confident. In 2005, the General Accounting Office found that the system's performance "remains uncertain and unverified, because MDA has not successfully completed a flight test using operationally representative hardware and software."[107] In subsequent years, the GAO found that although the National Missile Defense program gained "unprecedented funding and decision-making flexibility," it repeatedly fell short of goals while exceeding budgets.[108] By 2008, the MDA had still not achieved its Block 2006 goals, yet it was already more than $1 billion over the budget for that block. Because the MDA continued to defer tests, congressional investigators could not assess progress.[109] Furthermore, because the program was exempted from normal procedures, the computerized command and control system did not meet the integrated requirements of diverse commands that would actually operate the systems.[110]

3   A Reset or Risks Renewed?

3.1   The Fallacy of "Proven" and "Adaptive" Defenses

Just two days after an election victory in November 2008, Barack Obama confronted Russian warnings that a deployment of defenses in Eastern Europe would be countered by a deployment of short-range missiles in

Western Russia.[111] Obama dared not back down—accusations of military weakness had dogged his historic election campaign. Nonetheless, during his first months in office, Obama offered to forego the European-based midcourse defense in exchange for Russian help in stopping the proliferation of missile and nuclear technology to Iran. President Medvedev welcomed the shift. Nonetheless, when Obama outlined a more specific plan in September 2009, he denied accusations that he was accommodating Russian concerns. Instead, Obama highlighted new intelligence estimates suggesting that Iran would likely develop intermediate-range missiles before ICBMs.[112]

These threats remain the primary justification for the "European Phased Adaptive Approach" to defense. An advanced version of the Aegis interceptor, the Standard Missile 3 (SM-3), is the cornerstone of the new approach, which aims to defend Europe from Iranian intermediate-range missiles, and to help defend the United States from Iranian ICBMs.

It seems that Obama's shift has enabled some diplomatic progress. In April 2010, Obama and Medvedev signed a new START treaty, agreeing to cut the number of strategic missile launchers in half.[113] Nonetheless, missile defense remains a sticking point in U.S.-Russian relations. Proponents of missile defense nearly blocked the Senate ratification of the new START treaty in 2010, arguing that it would limit defenses because its preamble recognized a relationship between strategic offenses and defenses.[114] Obama and others continue to insist that the treaty places no limits on defenses. But many Russians disagree. The Duma (the lower house in Russia's parliament) ratified the treaty only after adding its own understanding that Russia might withdraw from the treaty if the United States deployed an unacceptable defense.[115]

Neither the preamble nor the Duma's memorandum of understanding hold legal force, but they continue to haunt negotiations on defensive cooperation and offensive reductions. In May 2011, Medvedev warned that Russia might be forced to withdraw from START if U.S. and NATO defenses did not give Russia a "finger on the trigger." Studies by Russia's Defense Ministry find that later phases of the SM-3 deployment, scheduled for the 2018 time frame, could threaten strategic parity.[116] U.S. physicists such as Ted Postol have confirmed these assessments—if all goes as planned, advanced versions of the SM-3 interceptors will be able to hit ICBMs as they descend into their terminal phase of flight.[117] Thus, the European system continues to risk undermining agreements to limit nuclear weapons.

Furthermore, though plans for defenses in Europe have shifted, most of the Missile Defense Agency's programs continue unchanged. In fact, the Obama administration claims the United States "is currently protected against the threat of limited ICBM attack" by the midcourse defense system.[118] The testing record suggests otherwise. As of May 2012, the MDA still only claims 8 successes out of 15 midcourse intercept tests, and most of these were highly artificial.[119] We can expect that the system will continue to struggle, because the satellite tracking and surveillance networks that might guide the EKV to its target had yet to be completed in 2011—in no small part because of complex software.[120] In the rush to deploy, the EKV design was based upon the myopic Mercad sensor, which has little hope of intercepting targets without guidance from the tracking satellites. We can only hope that Obama's publicly proclaimed confidence in the deployed National Missile Defense is aimed at its domestic supporters, for it would be dangerous for any policymaker to take a military or diplomatic gamble that the United States is protected from even the most limited of ICBM attacks.

Similar dangers lie in the new European approach. Obama has underscored a commitment to "deploy technology that is proven, cost-effective, and adaptable."[121] Yet, even the simplest variant of the SM-3 was only deployed after Obama spoke, and it has yet to see use in a real-world battle. All of the "successful" tests have relied upon interceptors approaching warheads broadside rather than head-on, and upon advance information about target length and rough location.[122] In 2011, a Defense Science Board panel concluded that the ground-based midcourse defense system had not yet solved the problems of discriminating warheads from decoys, and that if the system were to start "shooting at missile junk or decoys," the impact on the defense "would be dramatic and devastating!"[123]

In a real engagement with an unknown target size, location, and orientation, there is no assurance that the SM-3 will intercept a warhead. Furthermore, talk of "proven" *and* "adaptable" defense furthers a dangerous fallacy—that defensive systems proven in one context *remain* proven as they are adapted to meet new threats, that they can be readily black-boxed and shipped around the world.

The United States cannot simply black-box its defensive systems, because each must interoperate with other systems, many under multinational control. For example, the United States, Italy ,and Germany are developing a jointly controlled Medium Extended Air Defense System (MEADS). While the SM-3 defenses target high-altitude missile threats, MEADS aims to use Patriot interceptors to defend Europe from low-altitude threats, such as cruise missiles and bombers.[124] The Missile Defense Agency also continues to develop the Army's Terminal High Altitude Area Defense (THAAD) as a "globally transportable, rapidly deployable" system that might eventually find a place in European defenses.[125] Meanwhile, European nations are developing a separate, NATO-controlled defense of European territory. France and other nations have objected to the dominance of U.S. industries and military commands in NATO's defensive system, and urged other countries to contribute their diverse technologies to the system.[126] NATO has committed to upgrading its defensive command and control system so that it will interoperate with the U.S. system in Europe—but crucially, these systems will remain distinct.[127]

The U.S. defensive deployments do not simply "plug and play" in this complex and rapidly changing world. Instead, many military commands must negotiate diverse requirements, engineers must carefully design computer systems that meet these requirements, and tests must ensure interoperability among these systems. To achieve "layered" defenses over a large range of territories, each defensive system must receive and/or pass target tracks along to others, and computers must "fuse" data to provide a variety of different commands with a seamless picture of the battle space. The GAO recently warned that because the European Phased Adaptive Approach to defense is not synchronizing the development of these diverse systems, it is increasing the risks that the system "will not meet the warfighter's needs, with significant potential cost and schedule growth consequences."[128]

Furthermore, none of the risks associated with command and control systems will ever go away. As new technologies are added or improved, requirements must be again negotiated, and software engineers must upgrade and modify the system. The problems encoded in an initial design will raise the barrier to later improvements. Defenses can never be "black-boxed"—they will always be part of an evolving socio-technical system.

3.2 No Silver Bullet

In short, missile defenses in Europe confront "arbitrary complexity"—precisely what prompted Fred Brooks to warn that there could be "no silver bullet" to end the challenges of software development. Today we are more familiar than ever with the risks of software and other complex technological systems—tragedies such as the 2011 meltdown at Japan's Fukushima nuclear power plant continue to echo around the world. As software has become increasingly central to safety-critical systems—not only defenses, but also nuclear reactors, transportation, and hospital systems—governments around the world have sought guarantees of safety.

Many regulators seek quantitative guarantees. But Nancy Leveson, a computer scientist studying safety-critical systems, distinguishes between quantitative measures of *reliability*, the probability that an error will not occur, and *safety*, the probability that a system will not cause a disaster. Software that operates exactly as designed can still cause "severe mishaps."[129] Drawing upon the disciplinary repertoire of software engineering, Leveson has shown that software only becomes "safe" in relation to a broader system that may include physical and social elements. As Leveson notes, such notions of safety "may not be as amenable to quantitative treatment as reliability."[130] Government studies of safety-critical software have echoed these findings for decades, emphasizing that quantification still can't guarantee reliability at the extremely high levels we would like for safety-critical systems.[131]

Regulators have also sought guarantees in professional licensing standards. Unfortunately, these too have proven problematic. For example, in 1998, the Texas Board of Professional Engineers sparked controversy when it announced that it would recognize "software engineering" as a discipline for licensing. Under Texas law, only individuals who pass the board's licensing test are now authorized to provide software services to the public. But when the Texas Engineering Board turned to the IEEE Computer Society and ACM for help with designing a test, it found no consensus on the knowledge required for software licensing. The IEEE Computer Society began certifying software development professionals in 2004, promising computer workers and their prospective employers a competitive edge.[132] By contrast, an ACM task force on licensing chaired by Leveson and one of her collaborators, John Knight, concluded in 2002 there was "no generally agreed upon comprehensive body of knowledge for software engineering of safety-critical systems."[133] Although licensing continues in various countries and locales, and many individuals pursue professional certification exams, these eclectic efforts are unlikely to have much effect on an increasingly global software industry. Thus, the ACM warns that licensing may only encourage a false sense of security.[134]

Unable to promise regulators quantitative guarantees or licensing standards, software engineers working on safety critical systems must resort to a patchwork of processes, such as careful specification and documentation of requirements, risk analysis, formal proofs, rigorous arguments, and ultimately testing.[135] But it is here that contemporary "software engineering" has the most to offer—for it is no longer the "black art" of the 1950s. At the 20th anniversary discussion of Brooks's "No Silver Bullet" paper in 2007, David Parnas urged software developers to quit hunting for silver and opt for well-known "lead bullets": the principles, techniques, and processes that require skills and hard work.[136]

Indeed, studies of Defense Department (DoD) software have consistently attributed problems to a failure to apply well-known methods. For example, Brooks's 1987 DSB task force on military software noted that software engineering methods had "dramatically improved over the last decade," but were "not generally practiced in DoD."[137] Similarly, though previous studies of military software had "provided an abundance of valid conclusions and detailed recommendations," most "remain unimplemented."[138] Worse, more than a decade later, another Defense Science Board task force repeated these findings. Reviewing Defense Department software in light of six studies of software completed since Brooks's report, the 2000 task force concluded that most previous recommendations had not been implemented, though they "remain valid and could significantly and positively impact DoD software development capability."[139]

Most studies urge the Defense Department to learn from the commercial world, which has achieved some instructive success.[140] By 1987, Brooks's panel noted that industry's success stemmed from an understanding that "requirements, and especially the user interface, require iterative development, with interspersed testing by users."[141] Variations of iterative development methods have proliferated since Brooks wrote, ranging from Barry Boehm's "spiral development" to "extreme programming," and are typically used before software is formally "implemented" or released for general use. Arguably, iterative methods have found their ultimate success in an expanding software market, as billions of customers are enlisted—many unwillingly—in ongoing operational testing. For example, with 1.5 billion computers running Windows operating systems by 2011, Microsoft can draw upon a vast database for developing software.[142] Apple sold 18.6 million iPhones in the first quarter of 2011 alone.[143] Even relatively small software companies can typically enroll large numbers of users in software development.

This testing "economy of scale" has proven so successful in the software industry that many studies have urged the Defense Department to use more "Commercial-Off-the-Shelf" (COTS) software.[144] Broader defense acquisition reforms have also encouraged the military to buy, rather than develop, more of its software in the 1990s—with considerable effect. For example, in 2007 over 42 percent of the Army's Future Combat Systems computer code came from COTS or open source software.[145] But the needs of the modern military cannot be entirely met through commercial software, which carries unique risks of cyberterrorism. And in any case, the commercial world has no silver bullet either. A 2010 National Academies of Science study emphasizes that industry cannot be expected "to innovate at a rate fast enough to solve the DoD's hard technical problems."[146] Noting that defense acquisition policies have discouraged the military from fully applying iterative incremental development—an-oft repeated finding—recent studies recommend developing distinct acquisition processes for software that recognize the uniqueness of information technology.[147]

These recommendations may have some effect in the current administration.[148] But even if acquisition policies eventually encourage iterative, incremental software development, there are fundamental reasons to expect that software for missile defense will not be trustworthy. Today's missile defense project managers cite the need for iterative, incremental development when they argue that defensive systems are so complex that they can only be fully developed and tested after they are deployed. But this argument misses the crucial reason that iterative, incremental development improves software: many users discover mistakes that inevitably emerge through operational experience. Iterative development methods can only approach "reliability" as companies enroll thousands or millions of users to test the product in real-world use. Industry studies suggest that users encounter the most common problems relatively quickly, but that the majority of faults emerge only slowly, as unusual circumstances appear.[149]

Similarly, many faults in missile defense systems have been discovered and fixed during peacetime operations. But the unpredictable nature of a real-world attack will open the door for new bugs to emerge. Though testing aims to anticipate and correct any problems in an attack, missile defense has begun deployment on the basis of only a few dozen tests, all under highly artificial conditions, and with relatively little success. The SM-3 missile systems have never seen use in battle, and were tested under conditions quite different from those they might encounter in the European theater. And the midcourse defense system has never even seen the initial operational test that is typically required prior to weapons deployment. It is not impossible that these systems could intercept an unexpected missile—but it is highly unlikely. As Coyle argued in 2004, National Missile Defense is "no more than a scarecrow, not a real defense."[150]

3.3 The Risks of a Scarecrow Defense

Some argue that a convincing scarecrow is quite useful—it might deter an enemy from attacking. But a persuasive defense can also create new risks. The highly visible investment in defenses may elevate the status of the Iranian and North Korean threats in the international community, encouraging weapons development among those who want to become part of the "nuclear club."

Furthermore, any defense capable of intercepting missiles in space can certainly threaten satellites—the precious eyes and ears of every modern military. Thus, any midcourse defense convincing enough to deter an attack will raise anxieties about vulnerable satellites. In January 2007, nations around the world protested China's test of an antisatellite weapon.[151] Similarly, when the SM-3 was modified and used to intercept an errant U.S. spy satellite as it veered into the atmosphere, Russia and China both expressed concern that the United States might soon threaten perfectly functional satellites in space. Missile defenses need not be capable of actually intercepting satellites to spark a conflict. If military leaders in Russia or China see a surveillance satellite signal disappear unexpectedly—as occasionally happens when a satellite malfunctions—they might fear that the United States has destroyed the satellite in preparation for an attack.[152] Such fears could encourage preemptive military action.

Defenses can also create new risks after a conflict has begun. When Patriot battalions were sent to Iraq in 2003, televised broadcasts once again featured brilliant interceptions. The Pentagon later claimed that the defenses had incapacitated nine out of nine missiles headed toward a defended asset.[153] But, on March 23, 2003, a Patriot battery stationed near the Kuwait border shot down a British Tornado fighter jet, killing both crew members. Just two days later, operators in another battery "locked on" to a target and prepared to fire, discovering that it was an American F-16 only after the fighter fired back. Fortunately for all, only the Patriot's radar dish was destroyed.[154] Yet, less than one week later, tragedy struck again when another Patriot battery shot down an American Navy Hornet fighter, killing its pilot.[155]

A Defense Science Board (DSB) task force eventually attributed the failure to computer-related problems. The Patriot's Identify Friend or Foe (IFF) algorithms, which ought to have clearly distinguished allies from enemies, performed poorly. Additionally, command and control systems did not give crews good situational awareness, leaving them completely dependent on the faulty IFF technologies. The Patriot's protocols, displays, and software made operations "largely automatic," while "operators were trained to trust the software."[156] Despite decades of development and operational experience with NATO allies, interoperability remained elusive, with the U.S. Army inadvertently shooting down friendly Navy and Air Force aircraft.

The Patriot is not the only defensive system that might shoot down airplanes. The SM-3 systems currently deployed in Europe will aim at short- to intermediate-range missiles—precisely the targets that the Patriot was seeking when it shot down fighters. Government auditors recently warned that deploying the SM-3 rapidly, before establishing requirements or completing tests for interoperability with the partnering systems controlled by NATO, puts the new deployment at risk of providing operators with poor situational awareness or a confusing image of the battle space.[157] Similar limitations on situational awareness were to blame for the downing of the Tornado and F-16 fighters in 2003. Furthermore, as the DSB noted, the Patriot friendly fire was "not exactly a surprise"—the IFF had not performed well in training exercises.[158] But when the Bush administration decided to go to war, the fallible technology was rushed into a new and unpredictable operating environment.

This suggests an additional risk of confidence in "proven" and "adaptable" defenses: it can encourage bold but dangerous military interventions. After the "video-game war" of 1991, the notion of a clean and easy war seemed more plausible to many Americans. Preparing to invade Iraq in the fall of 2002, Rumsfeld ignored the advice of seasoned military planners, ordering commanders to rewrite their war plans to capitalize on precision weapons and reduce the number of ground troops to be deployed.[159] The Bush administration launched air strikes on Baghdad on March 22, 2003, televising a campaign of "shock and awe" around the world. On May 1, 2003, Bush declared "mission accomplished," and promised that "with new tactics and precision weapons, we can achieve military objectives without directing violence against civilians."[160]

As we now know, this confidence was tragically misplaced. Many defense analysts have argued that Rumsfeld's early reliance on high technology and underestimate of the troops needed in Iraq likely increased American casualties and prolonged the conflict.[161] The complexities of politics in Iraq continue to thwart stability. The limits of high technology should be all too clear.

3.4 Technological Progress

Today we have ample evidence of the risks of complex software, as well as complex technology more generally. Why, then, do we continue to invest in and rely upon complex technological systems—not just missile defense, but also nuclear reactors, massive oil production and distribution systems, and labyrinthine financial infrastructures?

As we have seen, the answers lie in intertwined cultural, political, and economic histories. Since the beginning of the nuclear era, the pursuit of defense has been driven by a deeply American belief in revolutionary technology. At the turn of the millennium, when confronted with a dismal testing record for the midcourse defense system, the director of the Missile Defense Agency, Donald Kadish, still appealed to nationalistic progress narratives, arguing that "birthing a revolutionary technology and making it useful is a tough engineering job that requires discipline, patience, and vision."[162] In congressional testimony, he rejected calls to slow the pace. Instead he told a story of "national leadership" that "persisted despite frustrations" with other weapons systems, thereby making "profound contributions to our national security."[163]

Similarly, when Kadish was challenged on the Patriot's friendly fire in 2003, he emphasized "absolutely tremendous" success in intercepting missiles.[164] The Patriot's real problem, he argued, was its performance in an "air defense mode."[165] These comments obscured the fact that the Patriot batteries had been searching for missiles, not aircraft, when they misidentified and shot down both fighters.[166] Nonetheless, celebratory videos and reports of the Patriot's successes in Iraq continue to circulate on the Web, while Raytheon, the Patriot manufacturer, boasts a "combat-proven" defense that has become the "system of choice for 12 nations around the globe."[167]

A persistent belief in technological progress all too often encourages experts to treat the challenges of complex software as "merely" social issues that can be resolved with good American engineering. In the wake of Bush's decision to deploy defenses, two computer science professors reiterated the challenges of complex software in *IEEE Technology and Society Magazine*, arguing that National Missile Defenses were unlikely to work properly when needed.[168] But one former industry programmer, Mark Halperin, disagreed in the *New Atlantis*. Underscoring technological progress, Halperin insisted that there is "nothing intrinsically problematic about building the necessary software" and that "we can have software that is virtually perfect if we are willing to pay for it."[169] Many of Halperin's assertions fly in the face of well-known problems in software engineering. For example, he states that software "is perfectly stable," contradicting decades of research on designing and managing software for constant change and reuse, including Parnas's seminal work in the 1970s.[170] Nonetheless Halperin's essay, published in a visible magazine, underscores the ways in which professional commitments are intertwined with nationalistic narratives of technological progress.

Confidence in technological progress also continues to support, and be supported by, political and economic institutions that change slowly. The risks of complex systems do not make good talking points in a world in which high technology symbolizes national prowess. Carefully paced testing schedules rarely match election cycles. Furthermore, the decision to deploy a missile defense system entailed a costly commitment to defense contractors and the engineers they employ, and it is politically risky to change course. Annual spending on missile defense nearly doubled to $7.8 billion in 2002, as Bush prepared for its deployment, and has continued to hover around $8 billion a year, costing taxpayers $8.5 billion in 2011.[171]

If complex missile defense systems truly reduced the risks confronted by all Americans, this might be a worthwhile investment. But, as I have argued, the prospects that defenses can provide trustworthy protection from a nuclear missile attack remain slim. Meanwhile, missile defense deployments create risks—of encouraging nuclear proliferation, crisis instability, friendly fire, and overconfident military interventions. At a time when poverty has reached the highest level in 50 years, and most of those afflicted do not reap the rewards of our national investment in the defense industry, we would do well to consider alternatives to managing risks.[172] As Charles Perrow argued in his 2007 analysis of American infrastructures, the best way to reduce vulnerabilities is not necessarily to bolster existing systems, but to recognize that our real vulnerabilities lie within those same complex socio-technical systems that we too easily trust.[173] The Port of Los Angeles, importing over 5 million poorly monitored containers from around the world each year, is far more vulnerable to nuclear attack than its open skies.

Missile defenses might deserve investment if they can directly facilitate reductions in nuclear and missile technologies—for example, by guarding against cheating on an arms control agreement. However, defenses have typically been treated as an alternative to negotiations that might be equally effective in limiting nuclear threats. To be sure, we have good reason to be skeptical of the tortuous negotiations with countries such as Iran and North Korea. With sharp divisions over the prospect of limiting nuclear offenses, many argue that defensive technology is the best or only way to reduce the risks of a nuclear missile attack. Yet, we have equally good reason to question whether complex defensive systems—which by their very nature must remain unproven until the moment of truth—can provide any more certain protection than an agreement to limit nuclear weapons. Furthermore, diplomatic agreements to limit nuclear weapons can reduce the risks that defensive systems cannot, such as terrorists attacking with suitcase nukes or dirty bombs. Indeed, if the United States were to reduce its own nuclear stockpiles, it might help reduce the prestige associated with nuclear weapons, thereby reducing incentives to acquire nuclear weapons in the first place.

4   Conclusion: Defense in a Risky World

As we have seen, our understanding of the risks and promise of missile defense derives not from self-evident technological realities, nor from socially constructed ideas about nuclear war, but from a history of expertise that is simultaneously technological and social. Physicists emerged as the dominant experts in national security after World War II. The disciplinary repertoire of physics has been handed down through generations of physicists, along with a sense of responsibility for managing the bomb. Defense Department insiders such as Hans Bethe and Richard Garwin developed the earliest public analyses of physical countermeasures to missile defense. Generations have picked these up and passed them on—from Kurt Gottfried and other UCS founders, to some of today's leading critics of missile defense, such as Lisbeth Gronlund, George Lewis, and David Wright. Physicists have long countered progress narratives by analyzing defenses as idealized systems that function perfectly, limited only by nature. For example, when the American Physical Society assessed prospects for boost-phase defenses in 2004, its analysis assumed that each developmental weapon would meet its stated goals, function perfectly, and that no time would be required for decision making—only nature would limit the defenses.[174] And Ted Postol, a supporter of some boost-phase defenses, mounts a persuasive critique when he argues that "improvements in technology cannot defeat the laws of physics."[175]

While physics provides an invaluable tool for assessing the prospects and limits of complex systems, it does not capture the equally important challenges associated with engineering, deploying, and operating complex systems. Over the past 50 years, these latter challenges have become a critical focus of computer and software engineering, fields that are inevitably social as well as technical. Though software experts have gained status in the Defense Department, they are not the most visible participants in contemporary debates on nuclear weapons. When it comes to policy, computer experts have long focused more on issues of database privacy and security. Significantly, a search of the *Risks Digest* for "Patriot" turns up far more discussion on the Patriot Act and its threat to civil liberties than on failures of missile defenses.

At the same time, software and computer experts are playing a growing role in a world that has become dependent upon computers for energy generation, transmission, and distribution, manufacturing, finance, health care, transportation, and other critical infrastructures. Amid continuing debate over whether the U.S. Congress should hold private companies to minimum security standards, computer experts at the Electronic Frontier Foundation and elsewhere argue that regulation opens the door to government abuses of civil liberties.[176] Meanwhile, the U.S. government is recruiting skilled hackers to wage cyber warfare. As a National Security Agency official noted at a 2011 hacker conference: "Today it's cyber warriors that we're looking for, not rocket scientists."[177]

Yet the growing importance of computers is only one reason that software experts today are far better placed in the Defense Department than they were in the early 1950s. As we have seen, computer software experts became better able to speak authoritatively as they developed a disciplinary repertoire. Whereas the earliest studies of the SAGE air defense portrayed programming as the easy part of the project, software emerged as the dominant computing challenge from the very first Defense Department studies of Star Wars. While 1950s programming was a "black art" with no academic status, today's practitioners can speak of widely accepted best practices and disciplines that are taught in universities and technical institutes around the world. Software fiascoes were "well remembered, but not well-documented experiences"[178] in the 1960s. Today, discussions of system disasters continue to be recorded in *Software Engineering Notes*, books, and other forums that aim to learn from failure. With the rise of a disciplinary repertoire came agreement among the most prominent software experts that there can be no silver bullet for the challenges of arbitrary complexity. Whereas computer experts were relatively marginal participants in the earliest studies of air defense, today they direct ongoing studies of the challenges facing Defense Department software.

In sum, the disciplinary repertoire of computing enabled software experts to speak more authoritatively about the risks of software. We can find parallels in other fields of technoscience, where experts struggle to anticipate and govern the promise and risks of new technology. For example, as climate scientists have developed methods for quantifying carbon emissions and global temperatures, and as they have established associated institutions such as the International Panel on Climate Change, they have come to speak with greater certainty and authority about the causes and risks of global warming. Similarly, biotechnology experts have sought to quantify the risks of terrorism and potential benefits of the industry.[179] These fields do not fit easily within traditional categories such as "professions" or "disciplines," yet as they have established forums for discussing and documenting challenges in their field, and as they have developed methods of quantification, they have been able to speak with more authority about the promise and risks of their technologies.

Although software engineers remain optimistic about progress in their field, the most prominent experts have agreed that software is limited by arbitrary complexity—the complexity of human institutions that change frequently and unpredictably. This conclusion carries significant implications for missile defense policy. Rather than rapidly fielding complex defensive systems, in hopes that they will eventually be operated cooperatively, we should put cooperation first. This is not just good diplomacy—it is essential to the technological effectiveness of any cooperative defense. The technical requirements for a complex hardware system can only be established after agreement on how the system is to operate. Complex hardware and software systems are developed and improved most rapidly when users get involved in testing from the very beginning. Modifying and adapting such complex systems to meet the requirements of new users, after they are initially deployed, create new complexities and risks that could be avoided by showing more patience in early development and testing.

Thus, a robust defense against nuclear weapons will, in the words of Ronald Reagan, "take years, probably decades, of effort on many fronts." It will also require us to recognize that the risks we face can only be partly addressed by the physical ingenuity of America's top scientists and engineers. Indeed, those who dismiss engineering challenges associated with defense as *merely* social miss the crucial point: complex technological systems, with all of their promise and risk, can never be *only* physical—they are simultaneously social and political to the core.

Notes

Introduction

1. Anne Barnard, "Behind a Patriot Battery," *Boston Globe*, April 2, 2003; Jonathan Weisman, "Patriot Missiles Seemingly Falter for Second Time; Glitch in Software Suspected," *Washington Post*, March 26, 2003; Ross Kerber, "War in Iraq," *Boston Globe*, April 16, 2003.

2. "Sighting with SAGE," *Newsweek*, July 7, 1958.

3. U.S. House of Representatives Committee on Appropriations, *Department of Defense Appropriations for 1964; Part 6, Research, Development, Test, and Evaluation*, May 6, 1963, 207.

4. Charles Perrow, *Normal Accidents: Living with High-Risk Technologies* (1984; repr., Princeton, NJ: Princeton University Press, 1999). See also Gene Rochlin, *Trapped in the Net: The Unanticipated Consequences of Computerization* (Princeton, NJ: Princeton University Press, 1998); Scott Sagan, *The Limits of Safety: Organizations, Accidents, and Nuclear Weapons* (Princeton, NJ: Princeton University Press, 1993); P. Neumann, *Computer-Related Risks* (Reading, MA: Addison-Wesley, 1995).

5. See Trevor Pinch, "Testing—One, Two, Three . . . Testing!": Toward a Sociology of Testing," *Science, Technology, & Human Values* 18, no. 1 (1993).

6. These pressures contributed directly to failures of the Patriot missile defense system in the 1991 Gulf War. See M. Blair, S. Obenski, and P. Bridickas, "Patriot Missile Defense: Software Problem Led to System Failure at Dhahran," (Washington, DC: U.S. General Accounting Office, Information Management and Technology Division, 1992).

7. Historian Theodore Porter writes of a similar "disciplinary objectivity." Theodore Porter, *Trust in Numbers* (Princeton, NJ: Princeton University Press, 1996).

8. See, for example, Bruno Latour, *Science in Action: How to Follow Scientists and Engineers through Society* (Cambridge, MA: Harvard University Press, 1987); Bruno Latour and Steve Woolgar, *Laboratory Life: The Social Construction of Scientific Facts* (London and Beverly Hills: Sage, 1979).; A. E. Clarke, J. H. Fujimura, and L. E. Kay, *The Right*

Tools for the Job: At Work in Twentieth-Century Life Sciences (Princeton, NJ: Princeton University Press, 1992); Peter Galison, *Image & Logic: A Material Culture of Microphysics* (Chicago: University of Chicago Press, 1997); Karin Knorr-Cetina, *Epistemic Cultures: How the Sciences Make Knowledge* (Cambridge, MA: Harvard University Press, 1999); John Downer, "When the Chick Hits the Fan: Representativeness and Reproducibility in Technological Tests," *Social Studies of Science* 37, no. 1 (2007).

9. Stephen Hilgartner, *Science on Stage: Expert Advice as Public Drama* (Stanford, CA: Stanford University Press, 2000).

10. Sociologists have similarly analyzed culture as a heterogeneous repertoire for framing problems and strategies of action. See A. Swidler, "Culture in Action: Symbols and Strategies," *American Sociological Review* 51, no. 2 (1986); P. DiMaggio, "Culture and Cognition," *Annual Reviews in Sociology* 23, no. 1 (1997); Doug McAdam, Sidney Tarrow, and Charles Tilly, *Dynamics of Contention* (New York: Cambridge University Press, 2001).

11. Public images did not necessarily capture the activities of physicists. For a view of the contrast between these two, see Rebecca Press Schwartz, "The Making of the History of The Atomic Bomb: Henry DeWolf Smyth and the Historiography of the Manhattan Project" (PhD diss., Princeton University, in progress as of 1997) and David Kaiser, "The Atomic Secret in Red Hands? Cold War Fears of Theoretical Physicists," *Representations* 90 (2005).

12. Einstein is quoted in many ways on this point; the sentiment is captured in an article as follows: "It can scarcely be denied that the supreme goal of all theory is to make the irreducible basic elements as simple and as few as possible without having to surrender the adequate representation of a single datum of experience." Albert Einstein, "On the Method of Theoretical Physics," *Philosophy of Science* 1, no. 2 (1934): 165.

13. Frederick Brooks, "No Silver Bullet: Essence and Accidents of Software Engineering," *IEEE Computer* 20, no. 4 (1987).

14. For some early findings of this study, see Rebecca Slayton, "Speaking as Scientists: Computer Professionals in the Star Wars Debate," *History and Technology* 19, no. 4 (2004); R. Slayton, "Discursive Choices: Boycotting Star Wars Between Science and Politics," *Social Studies of Science* 37, no. 1 (2007).

15. Garwin, e-mail to author, June 27, 2007.

16. Ibid.

17. Senate Foreign Relations Committee, "Nuclear Test Ban Treaty," in *Senate Foreign Relations* (1963), 763.

18. Lawrence Freedman, *The Evolution of Nuclear Strategy*, 3rd ed. (New York: St. Martin's Press, 2003); Gregg Herken, *Counsels of War* (New York: Alfred A Knopf, 1985);

Gregg Herken, *Cardinal Choices: Presidential Science Advising from the Atomic Bomb to SDI* (New York: Oxford University Press, 1992); Fred Kaplan, *The Wizards of Armageddon* (New York: Simon & Schuster, 1983); Donald Baucom, *The Origins of SDI, 1944–1983* (Lawrence: University Press of Kansas, 1992).

19. See, for example, Emanuel Adler, "The Emergence of Cooperation," *International Organization* 42, no. 1 (1992); Steve Weber, "Realism, Detente, and Nuclear Weapons," *International Organization* 44, no. 1 (1990). Other studies acknowledge controversy but focus on bureaucratic politics as a decisive factor, for example, Eric Pratt, *Selling Strategic Defense: Interests, Ideologies, and the Arms Race* (Boulder, CO: Lynne Rienner Publishers, 1990); Erik K. Pratt, "Missile Defense Sponsors: Shifting Political Support for Strategic Defense After Reagan," *Asian Perspective* 25, no. 1 (2001); Ernest J. Yanarella, *The Missile Defense Controversy; Technology in Search of a Mission* (Lexington: University Press of Kentucky, 2002); Morton Halperin, "The Decision to Deploy the ABM: Bureaucratic and Domestic Politics in the Johnson Administration" *World Politics* 25, no. 1 (1972).

20. William Broad, *Teller's War: The Top Secret Story behind the Star Wars Deception* (New York: Simon & Schuster, 1992); Francis Fitzgerald, *Way Out There in the Blue: Reagan, Star Wars, and the End of the Cold War* (New York: Simon & Schuster, 2000). Kaplan, *The Wizards of Armageddon.*

21. This is especially true of accounts that take a particular position on missile defense. For example, see Harold Brown, *The Strategic Defense Initiative: Shield or Snare?* (Boulder, CO: Westview Press, 1987); Sanford Lakoff and Herbert York, *A Shield in Space? Technology, Politics, and the Strategic Defense Initiative* (Berkeley: University of California Press, 1988); Sidney D. Drell, Philip J. Farley, and David Holloway, "The Reagan Strategic Defense Initiative: A Technical Political, and Arms Control Assessment," (Stanford, CA: Stanford Center for International Security and Arms Control, 1984); Gerald Steinberg, ed., *Lost in Space: The Domestic Politics of the Strategic Defense Initiative* (Lexington, KY: Lexington Books, 1988).

22. See Gabrielle Hecht, "Planning a Technological Nation: Systems Thinking and the Politics of National Identity in Postwar France," in *Systems, Experts, and Computers: The Systems Approach in Management and Engineering, World War II and After*, ed. Agatha Hughes and Thomas P. Hughes (Cambridge, MA: MIT Press, 2000); Hilgartner, *Science on Stage*; Sheila Jasanoff, *The Fifth Branch* (Cambridge, MA: Harvard University Press, 1990); Thomas Gieryn, "Boundary-Work and the Demarcation of Science from Non-science: Strains and Interests in Professional Ideologies of Scientists," *American Sociological Review* 48(1983); David H. Guston, "Stabilizing the Boundary between US Politics and Science: The Role of the Office of Technology Transfer as a Boundary Organization," *Social Studies of Science* 29, no. 1 (1999). For a (rare) account of missile defense that emphasizes the interconnections of scientific advising and political decision making, see Columba Peoples, *Justifying Ballistic Missile Defense* (Cambridge: Cambridge University Press, 2010).

23. Gabrielle Hecht, *The Radiance of France: Nuclear Power and National Identity after World War II* (Cambridge, MA: MIT Press, 1998).

24. Ashton B. Carter, John D. Steinbruner, and Charles A. Zraket, eds., *Managing Nuclear Operations* (Washington, DC: The Brookings Institution, 1987); Paul Bracken, *The Command and Control of Nuclear Forces* (New Haven, CT: Yale University Press, 1983); B. G. Blair, *Strategic Command and Control: Redefining the Nuclear Threat* (Washington, DC: Brookings Institution Press, 1985). Sagan, *The Limits of Safety: Organizations, Accidents, and Nuclear Weapons.*

25. See for example Matthew Evangelista, *Innovation and the Arms Race: How the United States and the Soviet Union Develop New Military Technologies* (Ithaca, NY: Cornell University Press, 1988). Ted Greenwood, *Making the MIRV: A Study of Defense Decision-Making* (Cambridge, MA: Ballinger Publishing Co., 1975).

26. For more on how organizational framing shapes defense decision making, see Lynn Eden, *Whole World on Fire: Organizations, Knowledge, and Nuclear Weapons Devastation* (Ithaca, NY: Cornell University Press, 2004).

27. "Remarks by the President on Strengthening Missile Defense in Europe," September 29, 2009. http://www.whitehouse.gov/the_press_office/Remarks-by-the-President -on-Strengthening-Missile-Defense-in-Europe.

28. See for example Lisbeth Gronlund et al., "Technical Realities: An Analysis of the 2004 Deployment of a U.S. National Missile Defense System," (Cambridge, MA: Union of Concerned Scientists, MIT Security Studies Program, 2004).

29. Trevor Pinch makes the very useful distinction between prospective and other kinds of testing: Pinch, "Testing—One, Two, Three . . . Testing!": Toward a Sociology of Testing." John Downer has also provided a useful analysis of the epistemic differences between these kinds of testing: "'737-Cabriolet': The Limits of Knowledge and the Sociology of Failure," *American Journal of Sociology* 117, no. 3 (2011): 725–762.

30. GAO, "Missile Defense: European Phased Adaptive Approach Acquisitions Face Synchronization, Transparency, and Accountability Challenges," (Washington, DC: GAO, 2010).

31. Paul N. Edwards, *The Closed World: Computers and the Politics of Discourse in Cold War America* (Cambridge, MA: MIT Press, 1996). Others have shown how computers-as-calculators (rather than real-time devices) shaped military planning. See, for example, Sharon Ghamari-Tabrizi, "Simulating the Unthinkable: Gaming Future War in the 1950's and 1960's," *Social Studies of Science* 30, no. 2 (2000); Sharon Ghamari-Tabrizi, *The Worlds of Herman Kahn: The Intuitive Science of Thermonuclear War* (Cambridge, MA: Harvard University Press, 2005).

32. Fred Turner, *From Counterculture to Cyberculture* (Chicago: University of Chicago Press, 2006). See also J. Markoff, *What the Dormouse Said: How the 60s Counterculture Shaped the Personal Computer* (New York: Viking, 2005).

33. Andrew Abbott, *The System of Professions: An Essay on the Division of Expert Labor* (Chicago: University of Chicago Press, 1988).

34. For more on this view, see Talcott Parsons, "The Professions and Social Structure," *Social Forces* 17 (1938); Don K. Price, *The Scientific Estate* (Cambridge, MA: Harvard University Press, 1965); Eliot Freidson, *Professionalism: The Third Logic* (Chicago: University of Chicago Press, 2001).

35. Nathan J. Ensmenger, *The Computer Boys Take Over: Computers, Programmers, and the Politics of Technical Expertise* (Cambridge, MA: MIT Press, 2010). See also Michael Mahoney, "Software as Science—Science as Software," in *History of Computing: Software Issues*, ed. Ulf Hashagen, Reinhard Keil-Slawik, and Arthur Norberg (Berlin: Springer-Verlag, 2000); Michael Mahoney, "Finding a History for Software Engineering," *IEEE Annals of the History of Computing* (2004).

36. Atsushi Akera, *Calculating a Natural World: Scientists, Engineers, and Computers During the Rise of U.S. Cold War Research* (Cambridge, MA: MIT Press, 2006).

37. Thomas Haigh, "Masculinity and the Machine Man," in *Gender Codes: Why Women Are Leaving Computing* (Los Alamitos, CA: Wiley-IEEE Computer Society Press, 2010).

38. Gieryn, "Boundary-Work and the Demarcation of Science from Non-science."

39. For foundational ethnographic work on the practices of science in laboratories, see Knorr-Cetina, *Epistemic Cultures*; for work emphasizing the practiced nature of theory and pedagogy, see David Kaiser, *Drawing Theories Apart: The Dispersion of Feynman Diagrams in Postwar Physics* (Chicago: University of Chicago Press, 2005).

40. I adopt Sheila Jasanoff's definition of political culture as the "systematic means by which a political community makes binding choices." Sheila Jasanoff, *Designs on Nature* (Princeton, NJ: Princeton University Press, 2005), 259.

41. For the strongest articulation of this view, see David Collingridge and Colin Reeve, *Science Speaks to Power: The Role of Experts in Policymaking* (New York: St. Martin's Press, 1986).

42. Some have suggested that STS scholars have reduced everything to politics, though a closer reading of the literature suggests otherwise. See the exchange in *Social Studies of Science*: H. M. Collins and Robert Evans, "The Third Wave of Science Studies: Studies of Expertise and Experience," *Social Studies of Science* 32, no. 2 (2002); H. M. Collins and Robert Evans, "King Canute Meets the Beach Boys: Responses to the Third Wave," *Social Studies of Science* 33, no. 3 (2003); Sheila Jasanoff, "Breaking the Waves in Science Studies: Comment on HM Collins and Robert Evans, 'The Third Wave of Science Studies,'" *Social Studies of Science* 33, no. 3 (2003); Arie Rip, "Constructing Expertise: In a Third Wave of Science Studies?," *Social Studies of Science*, no. 3 (2003); Brian Wynne, "Seasick on the Third Wave? Subverting the Hegemony of Propositionalism: Response to Collins & Evans (2002)," *Social Studies of Science* 33, no. 3 (2003).

43. See, for example, Hilgartner, *Science on Stage: Expert Advice as Public Drama*. Sheila Jasanoff, *The Fifth Branch: Science Advisors as Policymakers* (Cambridge, MA: Harvard University Press, 1990).

44. Steven Epstein, *Impure Science: Aids, Activism, and the Politics of Knowledge* (Los Angeles: University of California Press, 1996); Kelly Moore, *Disrupting Science: Social Movements, American Scientists, and the Politics of the Military, 1945–1975* (Princeton: Princeton University Press, 2008); Jasanoff, "Breaking the Waves in Science Studies."

45. My analysis of policymaking borrows from John Kingdon, James March, and others who suggest that decisions are made through unexpected alignments between four relatively unrelated streams running through organizations: problems, solutions, people, and windows of opportunity. John W Kingdon, *Agendas, Alternatives, and Public Policies* (Boston: Little, Brown and Company, 1984); M. D. Cohen, J. G. March, and J. P. Olsen, "A Garbage Can Model of Organizational Choice," *Administrative Science Quarterly* (1972).

46. Thomas Hughes, *American Genesis* (New York: Penguin Books, 1989).

47. Jasanoff, *Designs on Nature*.

48. For past discussions of the SAGE software problems, see Kent C. Redmond and Thomas M. Smith, *From Whirlwind to MITRE: The R&D Story of the SAGE Air Defense Computer* (Cambridge, MA: MIT Press, 2000); Claude Baum, *The System Builders: The Story of SDC* (Santa Monica, CA: System Development Corp., 1981); Edwards, *The Closed World*; Thomas P. Hughes, *Rescuing Prometheus* (New York: Pantheon, 1998).

49. For past accounts of this period, see Richard D. Burns and Lester H. Brune, *The Quest for Missile Defenses, 1944–2003* (Claremont, CA: Regina Books, 2003); Freedman, *The Evolution of Nuclear Strategy*; Herken, *Counsels of War*; Kaplan, *The Wizards of Armageddon*; Yanarella, *The Missile Defense Controversy*.

50. Arthur L. Norberg and Judy E. O'Neill, in *Transforming Computer Technology: Information Processing for the Pentagon, 1962–1986*, ed. Merritt Roe Smith (Baltimore: Johns Hopkins University Press, 1996); M. Mitchell Waldrop, *The Dream Machine: J.C.R. Licklider and the Revolution that Made Computing Personal* (New York: Viking Penguin, 2001).

51. Adler, "The Emergence of Cooperation"; Weber, "Realism, Detente, and Nuclear Weapons."

52. On the first software engineering conferences see Mahoney, "Finding a History for Software Engineering"; Ensmenger, *The Computer Boys Take Over: Computers, Programmers, and the Politics of Technical Expertise*; Ulf Hashagen, Reinhard Keil-Slawik, and Arthur L. Norberg, eds., *History of Computing: Software Issues* (Paderborn, Germany: Springer, 2000).

53. Geof Bowker, "How to Be Universal: Some Cybernetic Strategies, 1943–70," *Social Studies of Science* 23, no. 1 (1993). For a somewhat different account of the origins of "information technology" and its experts, see Ron Kline, "Cybernetics, Management Science, and Technology Policy: The Emergence of Information Technology as a Keyword," *Technology and Culture* 47, July (2006).

54. For studies that focus on physicists see Baucom, *The Origins of SDI, 1944–1983*; Broad, *Teller's War*; Fitzgerald, *Way Out There in the Blue*; Gordon R. Mitchell, *Strategic Deception: Rhetoric, Science, and Politics in Missile Defense Advocacy* (East Lansing: Michigan State University Press, 2000). For a different view, see Slayton, "Speaking as Scientists."

Chapter 1

1. John Jacobs, quoted in Henry S. Tropp, "A Perspective on SAGE: Discussion," *Annals of the History of Computing* (1983): 386.

2. George Valley, "How the SAGE Development Began," *IEEE Annals of the History of Computing* 7, no. 3 (1985): 197.

3. Ibid., p 205.

4. Ibid., p 207; Atsushi Akera, *Calculating a Natural World: Scientists, Engineers, and Computers During the Rise of U.S. Cold War Research* (Cambridge, MA: MIT Press, 2006).

5. Robert Everett, quoted in Herbert Benington, "The Production of Large Computer Systems," *IEEE Annals of the History of Computing* 5, no. 4 (1983): 350.

6. Kent C. Redmond and Thomas M. Smith, *From Whirlwind to MITRE: The R&D Story of the SAGE Air Defense Computer* (Cambridge, MA: MIT Press, 2000); Claude Baum, *The System Builders: The Story of SDC* (Santa Monica, CA: System Development Corp., 1981); Thomas P. Hughes, *Rescuing Prometheus* (New York: Pantheon, 1998).

7. Larry Owens, "The Counterproductive Management of Science in the Second World War: Vannevar Bush and the Office of Scientific Research and Development," *Business History Review* 68 (1994); David A. Grier, *When Computers Were Human* (Princeton, NJ: Princeton University Press, 2005); Jennifer Light, "When Computers Were Women," *Technology and Culture* 40, no. 3 (1999).

8. This chapter draws on several excellent accounts of the ENIAC. See Herman H. Goldstine, *The Computer from Pascal to von Neumann* (Princeton, NJ: Princeton University Press, 1972), 182; William Aspray, *John von Neumann and the Origins of Modern Computing* (Cambridge, MA: MIT Press, 1990), 26–27.

9. Quoted in Aspray, *John von Neumann and the Origins of Modern Computing*, 26–27.

10. Light, "When Computers Were Women," 470.

11. Aspray, *John von Neumann and the Origins of Modern Computing*, 26–27.

12. John Von Neumann and Herman H. Goldstine, *Planning and Coding of Problems for an Electronic Computer: Part II, Vol. I.* (Princeton, NJ: Institute for Advanced Study, 1947), 1–2. See also Goldstine, *The Computer from Pascal to von Neumann*, 269.

13. Some groups divided the process of programming and coding. See for example Grace Murray Hopper, "Automatic Programming—Definitions" (paper presented at the Symposium on Automatic Programming for Digital Computers, Washington, DC, May 13–14, 1954).

14. M. V. Wilkes, *Memoirs of a Computer Pioneer* (Cambridge, MA: MIT Press, 1985), 145.

15. M. V. Wilkes, D. J. Wheeler, and S. Gill, *The Preparation of Programs for an Electronic Digital Computer* (Cambridge, MA: Addison-Wesley, 1951).

16. Jay W. Forrester, "Digital Computers: Present and Future Trends," in *Proceedings of the Joint Computer Conference: Review of Electronic Digital Computers* (Philadelphia, PA, 1951).

17. Daniel Kevles, *The Physicists: The History of a Scientific Community in Modern America*, 2nd ed. (Cambridge, MA: Harvard University Press, 1987), 287–323; Christophe Lécuyer, "The Making of a Science Based Technological University: Karl Compton, James Killian, and the Reform of MIT, 1930–1957," *Historical Studies in the Physical and Biological Sciences* 23, no. 1 (1992); Peter Galison, *Image & Logic: A Material Culture of Microphysics* (Chicago: University of Chicago Press, 1997); Stuart W. Leslie, *The Cold War and American Science: The Military-Industrial-Academic Complex at MIT and Stanford* (New York: Columbia University Press, 1993).

18. James Killian, *The Education of a College President* (Cambridge, MA: MIT Press, 1985).

19. For more on these changes, see ibid.; Lécuyer, "The Making of a Science Based Technological University"; Silvan Schweber, "Big Science in Context: Cornell and MIT," in *Big Science: The Growth of Large-Scale Research*, ed. Peter Galison and Bruce Hevly (Stanford, CA: Stanford University Press, 1992); Silvan Schweber, "The Mutual Embrace of Science and the Military: ONR and the Growth of Physics in the United States After World War II," in *Science, Technology and the Military*, ed. Everett Mendelsohn, Merritt Roe Smith, and Peter Weingart (Boston: Kluwer Academic Publishers, 1988).

20. For more about the Rad Lab's systems engineering, see David Mindell, *Between Human and Machine: Feedback, Control, and Computing before Cybernetics* (Baltimore, MD: The Johns Hopkins University Press, 2002). For more biographical details on these individuals, see Ivan Getting, *All in a Lifetime: Science in the Defense of Democracy* (New York: Vintage Press, 1989); Ivan Getting, interview by Frederik Nebeker,

June 11, 1991, Boston, MA, transcript at http://www.ieeeghn.org/wiki/index.php/
Oral-History:Ivan_A._Getting_%281991%29; George Valley, interview by Andrew
Goldstein, June 13, 1991, Concord, MA, summary online at http://www.ieeeghn
.org/wiki/index.php/Oral-History:George_E._Valley.

21. Thomas A. Sturm, "The USAF Scientific Advisory Board," in *Special Studies*, ed.
Office of Air Force History (Washington, DC: USAF Historical Division Liasion
Office, 1967), 27–36.

22. Valley, "How the SAGE Development Began," 198–199.

23. See Kenneth Schaffel, *The Emerging Shield: The Air Force and the Evolution of Continental Air Defense 1945–1960* (Washington, DC: Office of Air Force History, U.S. Air
Force, 1991), 71, 87–89.

24. Valley, "How the SAGE Development Began," 199–200.

25. Ibid.

26. For example, see David Israel, "The Application of a High-Speed Digital Computer to the Present Day Air Traffic Control System," 8 June 1950, PW Dome, 63.

27. Edwards, *The Closed World*, 81; Akera, *Calculating a Natural World*, 205–211.

28. See, for example, discussion in Warren K. Lewis et al., "Report of the Committee
on Educational Survey," (Cambridge, MA: Massachusetts Institute of Technology,
1949), 58–59. See also Rebecca Slayton, "From a 'Dead Albatross' to Lincoln Labs:
Applied Research and the Making of a 'Normal' Cold War University," *Historical
Studies in the Natural Sciences* 42, no. 4 (2012): 255–282.

29. Julius Stratton to James Killian, 3 February 1950, AC 132, Box 14, Folder Project
Whirlwind. See also Redmond and Smith, *From Whirlwind to MITRE*, 77.

30. Valley, "How the SAGE Development Began," 211.

31. Ibid.

32. Stratton to Killian, 20 October 1950, AC 4, Box 211, Folder 3.

33. Valley, "How the SAGE Development Began," 211.

34. Stratton to Killian, 20 October 1951, AC 4, Box 211, Folder 3.

35. For discussions of an advisory group at this time, see Herken, *Cardinal Choices:
Presidential Science Advising from the Atomic Bomb to SDI* (New York: Oxford University Press, 1992).

36. Stratton to Killian, 2 January 1951, AC132, Box 14, Folder Project Charles.

37. Stratton to Killian, 2 January 1951, AC132, Box 14, Folder Project Charles.

38. See Redmond and Smith, *From Whirlwind to MITRE*, 98–108.

39. Vannevar Bush to William Webster, 19 March 1951, Bush Papers, Box 96, Folder Research and Development Board (September 1950–March 1951).

40. Redmond and Smith, *From Whirlwind to MITRE*, 91–93.

41. F. W. Loomis, *Problems of Air Defense: Final Report of Project Charles* (Cambridge, MA: MIT, 1951), 112–113.

42. Ibid., 117.

43. Robert Everett, "Whirlwind I," in *Proceedings of the Joint Computer Conference: Review of Electronic Digital Computers* (Philadelphia, PA: 1951).

44. Ibid.

45. Forrester, "Digital Computers: Present and Future Trends," 109–110.

46. Loomis, *Problems of Air Defense*, v–vi.

47. For an account of where the name SAGE comes from, see Valley, "How the SAGE Development Began."

48. Loomis, *Problems of Air Defense*, 117.

49. Quoted in Redmond and Smith, *From Whirlwind to MITRE*, 131.

50. Forrester and Everett to Lincoln Steering Committee, "Digital Computers for Air Defense System," 5 October 1951, PW Dome. See also ibid.

51. Loomis, "Problems of Air Defense: Final Report of Project Charles," xxi.

52. Redmond and Smith, *From Whirlwind to MITRE*, 113–116.

53. Valley to Killian, 10 April 1952, AC4, Box 135, Folder Lincoln.

54. For accounts of the summer study, see Allan Needell, *Science, Cold War, and the American State: Lloyd V Berkner and the Balance of Professional Ideals* (Amsterdam: Harwood Academic, 2000), 83–89; J. T. Jockel, *No Boundaries Upstairs: Canada, the United States, and the Origins of North American Air Defense, 1945–1958* (University of British Columbia Press, 1987), 65–66; Gregg Herken, *Counsels of War* (New York: Alfred A Knopf, 1985), 49; Fred Kaplan, *The Wizards of Armageddon* (New York: Simon & Schuster, 1983). Most historians attribute this phrase to NSC 68 (National Security Council, "United States Objectives and Programs for National Security," 1950); however, the phrase does not appear there, and its origins remain somewhat mysterious.

55. U.S. Strategic Bombing Survey (USSBS), "The Effects of Atomic Bombs on Hiroshima and Nagasaki," (Washington, DC: GPO, 1946), 43.

56. For more on Nitze, see N. Thompson, *The Hawk and the Dove: Paul Nitze, George Kennan, and the History of the Cold War* (Henry Holt and Co., 2009); Paul Nitze, *From Hiroshima to Glasnost* (New York: Grove Weidenfeld, 1989); Strobe Talbott, *The Master of the Game: Paul Nitze and the Nuclear Peace* (New York: Alfred A. Knopf, 1988).

57. Everett S. Gleason, "An Early Warning System" (NSC 139), 31 December 1952, DNSA, PD00311.

58. Herken, *Counsels of War*, 63.

59. Needell, *Science, Cold War, and the American State*; Jockel, *No Boundaries Upstairs*.

60. National Security Council, "Continental Defense" (NSC 5408), 11 February 1954, DNSA PD00388.

61. V. L. Adams, *Eisenhower's Fine Group of Fellows: Crafting a National Security Policy to Uphold the Great Equation* (Lanham MD: Lexington Books, 2006); Jockel, *No Boundaries Upstairs*.

62. Redmond and Smith, *From Whirlwind to MITRE*; Schaffel, *The Emerging Shield*.

63. "Group Leaders' Meeting," 24 November 1952, PW Dome.

64. "Group Leaders' Meeting," 14 July 1952, PW Dome.

65. Valley, "How the SAGE Development Began."

66. Redmond and Smith, *From Whirlwind to MITRE*; Schaffel, *The Emerging Shield*.

67. Redmond and Smith, *From Whirlwind to MITRE*, 278.

68. "Group 61 Air Defense Bi-weekly," 10 October 1952, PW Dome.

69. "Group Leaders' Meeting," 27 July 1953, PW Dome.

70. Redmond and Smith, *From Whirlwind to MITRE*, 315.

71. Forrester to Lincoln Steering Committee, "Organization and Tasks of Division 6," 16 November 1954, PW Dome.

72. Forrester, Everett, and Wieser to Lincoln Steering Committee, "Proposal for Accomplishing the SAGE System Computer Programming Tasks Outlined in Memorandum 6M-3416," 11 March 1955, PW Dome.

73. Ibid.

74. Ibid.

75. "Group Leaders' Meeting," 28 March 1955, PW Dome.

76. John Jacobs, *The SAGE Air Defense System: A Personal History* (Bedford, MA: MITRE Corporation, 1986), 94–95, 105.

77. Ibid.

78. Robert Crago in Tropp, "A Perspective on SAGE: Discussion," 386.

79. "SAGE System Meeting," 23 May 1955, PW Dome. See also Baum, *The System Builders: The Story of SDC*, 23. Redmond and Smith, *From Whirlwind to MITRE*, 376. Jacobs, *The SAGE Air Defense System*, 105.

80. Jacobs, *The SAGE Air Defense System*, 106.

81. "Group leaders' Meeting," 17 October 1955, PW Dome.

82. D. Kaiser, "The Physics of Spin: Sputnik Politics and American Physicists in the 1950s," *Social Research: An International Quarterly* 73, no. 4 (2006); David Kaiser, "Cold War Requisitions, Scientific Manpower, and the Production of American Physicists after World War II," *Historical Studies in the Physical and Biological Sciences* 33 (2002).

83. Tropp, "A Perspective on SAGE: Discussion," 386.

84. Jacobs, *The SAGE Air Defense System*, 106.

85. Tropp, "A Perspective on SAGE: Discussion," 386.

86. For a discussion of programming work and gendered divisions of labor, see Haigh, "Masculinity and the Machine Man."

87. Janet Abbate, "The Pleasure Paradox," in *Gender Codes: Why Women Are Leaving Computing*, ed. Thomas Misa (Hoboken, NJ: Wiley-IEEE Computer Society Press, 2010).

88. "Group Leaders' Meeting," 21 March 1955, PW Dome.

89. "Group Organization List," 1 February 1956, PW Dome.

90. Quoted in Redmond and Smith, *From Whirlwind to MITRE*, 376–378.

91. "Group Leaders' Meeting," 23 January 1956; "Sage System Meeting," 13 February 1956; "Group Leaders' Meeting," 20 February 1956, PW Dome.

92. "Group Leaders' Meeting," 20 February 1956, PW Dome.

93. Baum, *The System Builders: The Story of SDC*: 51. "SAGE System Meeting," 27 February 1956; "Group Leaders' Meeting," 20 August 1956, PW Dome.

94. "Group Leaders' Meeting," 9 April 1956, PW Dome.

95. "Group Leaders' Meeting," 28 May 1956, PW Dome.

96. Redmond and Smith, *From Whirlwind to MITRE*, 381.

97. Gen E. E. Partridge, "Further Emphasis on SAGE," 17 December 1956, in Richard F McMullen, "The Birth of SAGE, 1951–1958 (ADC Historical Study No. 33), Vol. 1," in *Air Defense Command Historical Studies* (Maxwell Air Force Base, Alabama: U.S. Air Force, 1965), 56–58.

98. Jacobs in Tropp, "A Perspective on SAGE: Discussion," 386.

99. Benington, "The Production of Large Computer Systems," 351.

100. Ibid.

101. Everett, editor's note in ibid., 350.

102. Ibid.

103. See Nathan J. Ensmenger, *The Computer Boys Take Over: Computers, Programmers, and the Politics of Technical Expertise* (Cambridge, MA: MIT Press, 2010).

104. Hopper, "Automatic Programming—Definitions."

105. Paul Armer, "SHARE—A Eulogy to Cooperative Effort," *Annals of the History of Computing* 2, no. 2 (1980).

106. A. Akera, "Voluntarism and the Fruits of Collaboration: The IBM User Group, SHARE," *Technology and Culture* 42, no. 4 (2001).

107. John Tukey, "The Teaching of Concrete Mathematics," *The American Mathematical Monthly* 65, no. 1 (1958): 2.

108. Thomas Haigh, "Software in the 1960s as Concept, Service, and Product," *IEEE Annals of the History of Computing* 24, no. 1 (2002).

109. Stephen B. Johnson, "Three Approaches to Big Technology: Operations Research, Systems Engineering, and Project Management," *Technology and Culture* 38, no. 4 (1997).

110. For example, see Lee A. DuBridge, "Scientific Manpower Problems," in *Proceedings of the February 4–6, 1953, Western Computer Conference* (Los Angeles, CA: ACM, 1953).

111. Quoted in Ensmenger, *The Computer Boys Take Over*, 91.

112. Quoted in Nathan J. Ensmenger, "Letting the 'Computer Boys' Take Over: Technology and the Politics of Organizational Transformation," *Int'l Review of Social History* 48 (2003): 162.

113. Abbate, "The Pleasure Paradox"; Ensmenger, *The Computer Boys Take Over*.

114. "Group Leaders' Meeting," 3 December 1956, PW Dome.

115. Baum, *The System Builders*, 47–51.

116. "Group Leaders' Meeting," 31 October 1955, PW Dome.

117. Redmond and Smith, *From Whirlwind to MITRE*, 407–408.

118. Everett to Valley, "Lincoln in SAGE," 22 June 1956. Quoted in ibid., 414.

119. Ibid.

120. Everett to Overhage, 3 June 1957, quoted in Redmond and Smith, 424.

121. James McCormack to Stratton, 5 April 1958, AC 134, Box 48, Folder 15.

122. Redmond and Smith, *From Whirlwind to MITRE*, 407–408.

123. Baum, *The System Builders*, 37.

124. Ibid., 38.

125. Ibid., 36.

Chapter 2

1. PSAC, "Report of the AICBM Panel," 21 May 1959, DNSA NH01357.

2. On science advising, refer to Gregg Herken's *Counsels of War* (New York: Alfred A Knopf, 1985); Gregg Herken, *Cardinal Choice: Presidential Science Advising from the Atomic Bomb to SDI* (New York: Oxford University Press, 1992). For accounts that focus on missile defense, see Donald Baucom, *The Origins of SDI, 1944–1983* (Lawrence, Kansas: University Press of Kansas, 1992); Richard D. Burns and Lester H. Brune, *The Quest for Missile Defenses, 1944–2003* (Claremont, CA: Regina Books, 2003); Ernest J. Yanarella, *The Missile Defense Controversy: Technology in Search of a Mission* (Lexington: University Press of Kentucky, 2002).

3. Baucom, *The Origins of SDI, 1944–1983*; Burns and Brune, *The Quest for Missile Defenses, 1944–2003*; Yanarella, *The Missile Defense Controversy*.

4. James Killian, *Sputnik, Scientists, and Eisenhower: A Memoir of the First Special Assistant to the President for Science and Technology* (Cambridge, MA: MIT Press, 1977), 66–67.

5. M. H. Armacost, *The Politics of Weapons Innovation: The Thor-Jupiter Controversy* (Columbia University Press, 1969), 57. Thomas P. Hughes, *Rescuing Prometheus* (New York: Pantheon, 1998); Davis Dyer, "Necessity as the Mother of Convention: Developing the ICBM, 1954–1958," *Business and Economic History* 22, no. 1 (1993).

6. Killian, *Sputnik, Scientists, and Eisenhower*, 68. Lee A. DuBridge, interview by Judith R. Goodstein, Pasadena, CA, February 19, 1981; Oral History Project, California Institute of Technology Archives, http://oralhistories.library.caltech.edu/68/1/OH _DuBridge_2.pdf.

7. V. L. Adams, *Eisenhower's Fine Group of Fellows: Crafting a National Security Policy to Uphold the Great Equation* (Lanham, MD: Lexington Books, 2006). For more on the panel, see Herken, *Counsels of War*, 106; Killian, *Sputnik, Scientists, and Eisenhower*, 67–92.

8. Joint Committee on Atomic Energy, *Missiles with Nuclear Warheads*, 84th Cong. 1rst. sess., May 25, 1955, 19.

9. For a short discussion, see Louis Brown, *A Radar History of World War II* (Bristol and Philadelphia: Institute of Physics, 1999), 292–300.

10. Although Killian's report does not describe countermeasures in detail, von Neumann's comments indicate that scientists and engineers were well aware of them.

11. "Comments on the Report to the President by the Technological Capabilities Panel of the Science Advisory Committee," with a cover memorandum by National Security Council Secretary James S. Lay, 8 June 1955, PD00460, DNSA.

12. H. W. Brands, *The Devil We Knew: Americans and the Cold War* (New York: Oxford University Press, 1993), 65.

13. Ibid.

14. Adams, *Eisenhower's Fine Group of Fellows*, 125.

15. "Group Leaders' meeting," 3 January 1955, PW Dome. For more on these early debates, see also Kenneth Schaffel, *The Emerging Shield* (chap. 1, n. 23), 235–236; Burns and Brune, *The Quest for Missile Defenses, 1944–2003*; Baucom, *The Origins of SDI, 1944–1983*.

16. Robert Frank Futrell, *Ideas, Concepts, Doctrine: A History of Basic Thinking in the United States Air Force, 1907–1964* (Maxwell Air Force Base, AL: Air University, 1971), 500–502.

17. Wiesner to Murphree (Special Assistant for Guided Missiles), 15 March 1957, MC420, Box 3, Folder Scientific Advisory Committee-AICBM Subcommittee.

18. Rear Admiral J. H. Sides, "Presentation to the National Security Council on the Anti-Ballistic Missile Program," 11 January 1957, DNSA, NH00568, 71. The label "Wizard" appears to have been applied to several different projects or programs involving unmanned interceptors of missiles or drones. This caused considerable confusion in Congressional hearings. See House Armed Services Committee, *Investigation of National Defense Missiles*, 85th Cong., 2nd sess., January 13, 1958.

19. Sides, "Presentation to the National Security Council on the Anti-Ballistic Missile Program," 71.

20. Ibid., 72.

21. Ibid., 72.

22. NSC 5707/4, "Review of National Security Policy: Military and Non-military Aspects of Continental Defense," 26 March 1957, DNSA, PD00506, 2.

23. Ibid.

24. NSC 5709, "A Federal Shelter Program for Civil Defense," 29 March 1957, DNSA, PD00514. See also Adams, *Eisenhower's Fine Group of Fellows*, 161.

25. Killian's files at MIT show that he consulted with Gaither in organizing Lincoln in the early 1950s. For more on the panel organization, see Adams, *Eisenhower's Fine Group of Fellows*, 160.

26. Quoted in ibid., 163–164.

27. Ibid.

28. Fred Kaplan, *The Wizards of Armageddon* (New York: Simon & Schuster, 1983), 117, 28–29.

29. Robert Cutler, *No Time for Rest* (1966), 354–355; D. L. Snead, *The Gaither Committee, Eisenhower, and the Cold War* (Columbus: Ohio State University Press, 1999).

30. Kenneth Kreps, "SAC and USAF Briefings for Steering Committee, Security Resources Panel," 24 September 1957, DNSA NH00093.

31. Ibid.

32. Ibid.

33. Quoted in Kaplan, *The Wizards of Armageddon*, 132–134, 433.

34. Ibid.

35. Security Resources Panel of the Science Advisory Committee, "Deterrence and Survival in the Nuclear Age," 7 November 1957, DNSA, NH01329, 29. Although Bradley's work is not outlined explicitly in the Gaither report, his contributions are described in Richard Barber, "History of DARPA" (Institute for Defense Analysis, 1975).

36. Security Resources Panel, "Deterrence and Survival," 8.

37. Ibid., 17–18.

38. Jerome B. Wiesner, *Where Science and Politics Meet* (New York: McGraw Hill, 1961), 174.

39. Security Resources Panel, "Deterrence and Survival," 17.

40. For a discussion of Paul Nitze's influence, see David Callahan, *Dangerous Capabilities: Paul Nitze and the Cold War* (New York: HarperCollins, 1990), 167–168; Adams, *Eisenhower's Fine Group of Fellows*, 174.

41. Security Resources Panel, "Deterrence and Survival," 1, 4.

42. Ibid., 8.

43. Ibid., 17.

44. Committee, "Deterrence and Survival," 17.

45. Ibid., 14.

46. Quoted in Herken, *Counsels of War*, 113, 16.

47. Gaither to Wiesner, 15 November 1957, MC420, Box 4, Folder SAC Study Group.

48. Chalmers Roberts, "Enormous Arms Outlay Is held Vital to Survival," *WP*, December 20, 1957. For a discussion of leaks, see Kaplan, *The Wizards of Armageddon*, 152–154; Snead, *The Gaither Committee, Eisenhower, and the Cold War*, 140.

49. E. P. Oliver, "Memorandum to the Security Resources Panel Steering Committee and Advisory Panel, Two Months Later," 14 January 1958, DNSA, NH00408.

50. "Memorandum of Conference with the President," 8 October 1957, DNSA NH00589.

51. Barber, "History of DARPA," II-5.

52. James Reston, "Army Plan Seeks 6 Billion to Make a Missile Killer," *New York Times*, November 20, 1957.

53. House Armed Services Committee, *Investigation of National Defense Missiles*, 4453.

54. Ibid., 4791; ibid., 4453.

55. Ibid., 4791; ibid., 4453.

56. Zuoyue Wang, *In Sputnik's Shadow: The President's Science Advisory Committee and Cold War America* (Rutgers University Press, 2008); Herken, *Cardinal Choices: Presidential Science Advising from the Atomic Bomb to SDI*: 102–106.

57. Barber, "History of DARPA," III-53–III-54.

58. James Killian, *The Education of a College President* (Cambridge, MA: MIT Press, 1985), 314.

59. Herbert York, *Making Weapons, Talking Peace: A Physicist's Odyssey from Hiroshima to Geneva* (New York: Basic Books, 1987); Barber, "History of DARPA," III-53–III-54.

60. House Science and Astronautics, *Missile Development and Space Sciences*, February 17–18, 1959, 330.

61. Barber, "History of DARPA," III-55, IV-24.

62. Fred A. Payne speech, October 1, 1964, quoted in Edward Randolph Jayne, "The ABM Debate: Strategic Defense and National Security" (PhD diss., Massachusetts Institute of Technology, 1969), 64. Barber's history of DARPA confirmed this view (Barber, "History of DARPA," III-50–III-51.

63. Barber, "History of DARPA," III-50–59.

64. Roy Johnson, "Memorandum for the Secretary of Defense: ARPA FY 1961 Budget Request," 3 August 1959, NARA, RG 359, Box 5, Folder DOD Budget 61.

65. House Appropriations Committee, *Department of Defense Appropriations for 1961, Part 6: Research, Development, Test and Evaluation*, 86th Cong. 2nd sess., March 9, 11, 14–18, 1960, 140–143.

66. The remainder of the budget in both 1959 and 1960 was dedicated to systems concepts, a category that became somewhat notorious; see discussion in Barber, "History of DARPA," V-19–V-22.

67. See, for example, House Appropriations Committee, *Department of Defense Appropriations for 1960. Part 6: Research, Development, and Testing*, 87th Cong. 1rst. sess. Apr. 14–16, 20–22, 24, 27, 1959, 111.

68. Sidney Reed, Richard H. Van Atta, and Seymour J. Deitchman, *DARPA Technical Accomplishments: An Historical Review of Selected DARPA Projects, Volume I* (Washington DC: Institute for Defense Analysis, 1990), 6-1; Barber, "History of DARPA," IV-26.

69. The first of these two issues is discussed in subsequent hearings on the test ban: Senate Foreign Relations Committee, *Nuclear Test Ban Treaty*, 88th Cong. 1rst sess. August 12–27, 1963. This final idea came from Nicholas Christofilos, a much-admired physicist at Livermore Laboratories. In January 1958, space scientist James van Allen confirmed that electrons were trapped in the earth's magnetic field in small numbers (hence the van Allen belts). For contemporary discussion, see William Trombley, "Triumph in Space for a 'Crazy Greek,'" *Life*, March 30, 1959.

70. General Goodpaster, "Memorandum of a Conference with the President," 30 October 1957, DNSA NH01327; Herken, *Cardinal Choices*, 103.

71. "Meeting of the Ad Hoc Panel on Nuclear Test Limitation," 15 March 1958, DNSA, NP00398.

72. Herken, *Cardinal Choices*; York, *Making Weapons, Talking Peace*, 118.

73. York, *Making Weapons, Talking Peace*, 118.

74. "Memorandum of a Conference with the President," 17 April 1958. DNSA, NH00623.

75. For objections, see Herbert Loper to Killian, 13 May 1958, "Analysis of a Report of the President's Science Advisory Committee on the Cessation of Nuclear Testing," DNSA, NH000059. See also Herken, *Cardinal Choices*, 110–113.

76. PSAC, "Report of the AICBM Panel," 4 November 1959, DNSA, NH01369, 4.

77. PSAC, "Report of the AICBM Panel," 4 November 1959, DNSA, NH01369, 3.

78. PSAC, "Report of the AICBM Panel," 21 May 1959, DNSA, NH01369, 6.

79. For discussion of the Kwajalein field tests, see Bell Laboratories, *ABM Research and Development at Bell Laboratories; Kwajalein Field Station* (Volume II of III) (Whippany, NJ: Bell Labs, on behalf of Western Electric, for U.S. Army Ballistic Missile Defense Systems Command, 1975), 27.

80. PSAC, "Report of the AICBM Panel," 21 May 1959, DNSA, NH01369, 9.

81. Wiesner, "Warning and Defense in the Missile Age," 3 June 1959, DNSA NH01361.

82. PSAC, "Nike-Zeus," 18 October 1960, DNSA NH01385, 2.

83. Ibid.

84. Ibid; Herbert York, *Race to Oblivion: A Participant's View of the Arms Race* (New York: Simon and Schuster, 1970), 141.

85. PSAC, "Report of an Ad Hoc Panel on Technical Aspects of the Nike-Zeus Pacific Test Program," 26 May 1960, DNSA NH01378.

86. York, *Making Weapons, Talking Peace: A Physicist's Odyssey from Hiroshima to Geneva*, 189.

87. York Papers, UCSD.

88. George Kistiakowsky, *A Scientist at the White House* (Cambridge, MA: Harvard University Press, 1976), 413.

89. Ibid.

90. PSAC, "Report of the AICBM Panel," May 1959, DNSA NH01357.

91. "Notes on Meeting in Cambridge, MA, Saturday, April 18, 1959 on Warning and Air Defense," DNSA, NH01355, 3.

92. See also "Report of the Early Warning Panel," March 13, 1959, in National Security Archives Electronic Briefing Book. http://www.gwu.edu/~nsarchiv/NSAEBB/NSAEBB235/03.pdf.

93. Albert Wohlstetter, Fred S. Hoffman, and Henry Rowen, "Protecting U.S. Power to Strike Back in the 1950s and 1960s (Staff Report R290)," (Santa Monica: Rand, 1956).

94. Ibid.

95. See discussion in Scott Sagan, *The Limits of Safety: Organizations, Accidents, and Nuclear Weapons* (Princeton, NJ: Princeton University Press, 1993), 165–170.

96. William J. Jorden, "Soviet Says SAC Flights Over Arctic Peril Peace," *New York Times*, April 19, 1958.

97. Frank H. Bartholomew, "This Is Article Cited by Soviet in Its Criticism of U.S. Flights," *New York Times*, April 19, 1958.

98. See discussion in Sagan, *The Limits of Safety: Organizations, Accidents, and Nuclear Weapons*, 167–170.

99. Thomas Schelling, *The Strategy of Conflict* (Cambridge, MA: Harvard University, 1960).

100. Herken, *Counsels of War*, 123. Bernhard Bechhoefer, *Postwar Negotiations for Arms Control* (Washington, DC: Brookings Institution, 1961), 476.

101. The novel was published in 1962; the movie came out in 1964.

102. *Dr. Strangelove* was based upon British novelist Peter George's *Two Hours to Doom*, published in 1958 in the United States as *Red Alert*.

103. "A-war by Accident?," *Newsweek*, March 24, 1958.

104. Norbert Wiener, "Some Moral and Technical Consequences of Automation," *Science* 131 (1960): 1357.

105. Wiesner, *Where Science and Politics Meet*, 202.

Chapter 3

1. House Appropriations, *Department of Defense Appropriations for 1964; Part 6, Research, Development, Test, and Evaluation*, May 6, 1963, 206–207.

2. Ibid., 233.

3. House Appropriations, *Department of Defense Appropriations for 1963; Part 5, Research, Development, Test, and Evaluation*, March 19, 20, 23, 1962, 114.

4. House Appropriations, *Department of Defense Appropriations for 1964; Part 6, Research, Development, Test, and Evaluation*, 233.

5. Ibid., 234.

6. Jerome Wiesner, "Communication Sciences in a University Environment," *IBM Journal* (1958): 268.

7. Ibid., 271.

8. Atsushi Akera, *Calculating a Natural World: Scientists, Engineers, and Computers during the Rise of U.S. Cold War Research*, Inside Technology (Cambridge, MA: MIT Press, 2006).

9. Arthur L. Norberg and Judy E. O'Neill, *Transforming Computer Technology: Information Processing for the Pentagon, 1962–1986*, ed. Merritt Roe Smith. Johns Hopkins Studies in the History of Technology (Baltimore: Johns Hopkins University Press, 1996).

10. Claude Baum, *The System Builders: The Story of SDC* (Santa Monica, CA: System Development Corp., 1981), 53.

11. MITRE, *MITRE: The First Twenty Years* (Bedford, MA: MITRE, 1979), 28–30.

12. See Paul N. Edwards, *The Closed World: Computers and the Politics of Discourse in Cold War America* (Cambridge, MA: MIT Press, 1996), 46.

13. MITRE, *MITRE: The First Twenty Years*.

14. Jack Raymond, "Kennedy's Secretary of Defense Faces Many Varied Problems," *New York Times*, December 14, 1960; Jack Raymond, "McNamara, Former Professor, Rose to the Top in Business," *New York Times*, December 14, 1960.

15. John F. Kennedy, "Recommendations Relating to Our Defense Budget; Document no. 123 for the House of Representatives, 87th Congress, 1rst Session" (House Committee on Appropriations, 1961), 8.

16. Herbert York, *Making Weapons, Talking Peace: A Physicist's Odyssey from Hiroshima to Geneva* (New York: Basic Books, 1987).

17. Fred Kaplan, *The Wizards of Armageddon* (New York: Simon & Schuster, 1983), 273.

18. Robert McNamara, Memorandum to the Chairman of the Joint Chiefs of Staff, "Command and Control," August 12, 1962, DNSA NH01275.

19. David E. Pearson, *The World Wide Military Command and Control System: Evolution and Effectiveness* (Maxwell Air Force Base, AL: Air University Press, 2000), 47–48.

20. J. S. Butz, "White House Command Post—1966," *Air Force Magazine* (1964), 76.

21. Pearson, *The World Wide Military Command and Control System.*

22. Editors, "Military Electronic Systems Catalog," *Armed Forces Management* 10, July (1964).

23. Butz, "White House Command Post—1966," 81.

24. See James Fallows, *National Defense* (New York: Random House, 1981), 59; Les Earnest, "The C3 Legacy, Part 3: Command-Control Catches On," *Risks Digest*, February 5, 1990. http://catless.ncl.ac.uk/Risks/9.67.html#subj4.1.

25. Frederick B. Thompson, "Fractionization of the Military Context" (paper presented at the Spring Joint Computer Conference, 1964).

26. J. S. Butz, "USAF and the Computer Revolution," *Air Force Magazine*, March (1964): 33.

27. Les Earnest, "The C3 Legacy, Part 4," *Risks Digest*, March 5, 1990. http://catless.ncl.ac.uk/Risks/9.74.html#subj3.1.

28. Ibid.

29. Les Earnest, "The C3 legacy, Part 5," Risks Digest, April 11, 1990. http://catless.ncl.ac.uk/Risks/9.80.html#subj8.1.

30. Les Earnest, "The C3 Legacy, Part 6," May 30, 1990. http://catless.ncl.ac.uk/Risks/9.97.html#subj1.1.

31. Gilbert Burck, "On Line in 'Real Time,'" *Fortune*, April 1964; Gilbert Burck, "The Boundless Age of the Computer," *Fortune*, March 1964.

32. D. G. Copeland, R. O. Mason, and J. L. McKenney, "SABRE: The Development of Information-Based Competence and Execution of Information-Based Competition," *Annals of the History of Computing, IEEE* 17, no. 3 (2002); Martin Campbell-Kelly, *From Airline Reservations to Sonic the Hedgehog* (Cambridge, MA: MIT Press, 2003).

33. Robert V. Head, "Getting Sabre Off the Ground," *IEEE Annals of the History of Computing* (2002): 37.

34. J. C. R. Licklider, "Underestimates and Overexpectations," in *ABM: An Evaluation of the Decision to Deploy an Antiballistic Missile Defense System*, ed. Abram Chayes and Jerome Wiesner (New York: Harper & Row, 1969).

35. "A Survey of Airline Reservation Systems," *Datamation*, June 1962: 54.; Copeland, Mason, and McKenney, "SABRE."

36. Thomas Haigh, "Inventing Information Systems: The Systems Men and the Computer, 1950–1968," *Business History Review* 75 (2001).

37. Ibid.; Nathan J. Ensmenger, "Letting the 'Computer Boys' Take Over: Technology and the Politics of Organizational Transformation," *Int'l Review of Social History* 48 (2003); Steve Barley, "Technicians in the Workplace: Ethnographic Evidence for Bringing Work into Organizational Studies," *Administrative Science Quarterly* 41, no. 3 (1996).

38. C. W. Borklund, "How to Succeed in Spite of Yourself," *Armed Forces Management* 10 (July 1964): 9.

39. "How Not to Build a C&C System Is Still Unanswered," *Armed Forces Management* 12 (July 1966): 109.

40. Ibid.

41. Editor, "Evolution and Compatibility: 1965's Key Words in Tactical C&C," *Armed Forces Management* 11 (July 1965): 44.

42. Hal Bamford, "Software is the Greater of the Two 'Ware' Problems," *Armed Forces Management* 9 (July 1963): 52–55.

43. See also J. A. N Lee, "Claims to the Term 'Time Sharing,'" *IEEE Annals of the History of Computing* 14, no. 1 (1992).

44. R. W. Bemer, "The Status of Automatic Programming for Scientific Computation," 107–117, and panel discussion, 118–126 in Proceedings of the 4th Annual Computer Applications Symposium, Armour Research Foundation, October 24–25, 1957. Quoted at http://www.bobbemer.com/TIMESHAR.HTM.

45. W. F. Bauer, "Computer Design from the Programmer's Viewpoint," in *Papers and Discussions Presented at the December 3–5, 1958, Eastern Joint Computer Conference: Modern Computers: Objectives, Designs, Applications* (Philadelphia, PA: ACM, 1958), 49.

46. Ibid.

47. Ibid.

48. Atsushi Akera, "Voluntarism and the Fruits of Collaboration: The IBM User Group, SHARE," *Technology and Culture* 42, no. 4 (2001).

49. D. Nofre, "Unraveling Algol: US, Europe, and the Creation of a Programming Language," *Annals of the History of Computing, IEEE* 32, no. 2 (2010); R. W. Bemer, "A Politico-Social History of Algol," in *Annual Review in Automatic Programming* 5, ed. Mark I. Halpern and Christopher J. Shaw (Oxford: Pergamon Press, 1969), 163.

50. Quoted in Jean Sammet, "The Early History of COBOL," *ACM SIGPLAN Notices* 13, no. 8 (1978): 124.

51. Charles A. Phillips, Joseph F. Cunningham, and John L .Jones, "Recollections on the Early Days of COBOL and CODASYL," *IEEE Annals of the History of Computing* 7, no. 4 (1985): 305.

52. Ben Shneiderman, "Recollections on the Early Days of COBOL and CODASYL," *IEEE Annals of the History of Computing* 7, no. 4 (1985): 351.

53. Baum, *The System Builders*, 55.

54. Nathan J. Ensmenger, *The Computer Boys Take Over* (Cambridge, MA: MIT Press, 2010).

55. A. M. Turing, "On Computable Numbers, with an Application to the Entscheidungsproblem," *Proceedings of the London Mathematical Society* 2, no. 1 (1937): 241. See also B. J. Copeland, ed. *The Essential Turing* (Oxford: Oxford University Press, 2004); B. J. Copeland and D. Anderson, "Alan Turing's Automatic Computing Engine: The Master Codebreaker's Struggle to Build The Modern Computer," *History and Philosophy of Logic* 29, no. 4 (2008).

56. Quoted in H. Crowther-Heyck, *Herbert A. Simon: The Bounds of Reason in Modern America* (Baltimore: Johns Hopkins University Press, 2005), 203. See also Baum, *The System Builders;* Sharon Ghamari-Tabrizi, "Simulating the Unthinkable: Gaming Future War in the 1950's and 1960's," *Social Studies of Science* 30, no. 2 (2000).

57. Crowther-Heyck, *Herbert A. Simon*; E. M. Sent, "Herbert A. Simon as a Cyborg Scientist," *Perspectives on Science* 8, no. 4 (2000).

58. H. A. Simon, *Models of My Life* (New York: Basic Books, 1991), 201.

59. Game theory was most famously articulated in *The Theory of Games and Economic Behavior*, coauthored with economist Oskar Morgenstern in 1944. C. E. Shannon, "XXII. Programming a Computer for Playing Chess," *Philosophical Magazine (Series 7)* 41, no. 314 (1950); A. M. Turing, "Computing Machinery and Intelligence," *Mind* 59, no. 236 (1950). For discussion of the relation between chess and artificial intelligence, see Nathan Ensmenger, "Is Chess the Drosophila of Artificial Intelligence? The Social History of an Algorithm," *Social Studies of Science* 42, no. 1 (2012).

60. Herbert Simon, "A Behavioral Model of Rational Choice (P-365)" (Santa Monica, CA: RAND, 1953). See also H. A. Simon, "A Behavioral Model of Rational Choice," *The Quarterly Journal of Economics* 69, no. 1 (1955), 2.

61. Crowther-Heyck, *Herbert A. Simon*.

62. Allen Newell, "The Chess Machine: An Example of Dealing with a Complex Task by Adaptation," in *Proceedings of the March 1–3, 1955, Western Joint Computer Conference* (Los Angeles, CA: ACM, 1955); A. Newell, "Description of the Air-Defense Experiments II: The Task Environment" (Santa Monica, CA: Rand, 1955).

63. Simon, *Models of My Life*, 206.

64. P. McCorduck, *Machines Who Think: A Personal Inquiry into the History and Prospects of Artificial Intelligence* (Natick, MA: AK Peters, Ltd., 2004).

65. Allen Newell, J. C. Shaw, and H. A. Simon, "Chess-Playing Programs and the Problem of Complexity," *IBM Journal of Research and Development* 2, no. 4 (1958), 119–128

66. Akera, *Calculating a Natural World*, 284–286.

67. Herbert Teager and John McCarthy, "Time-Shared Program Testing," in *Preprints of Papers Presented at the 14th National Meeting of the Association for Computing Machinery* (Cambridge, MA: ACM, 1959). "The Beginning at MIT," *IEEE Annals of the History of Computing*, vol. 14, no. 1 (January–March 1992): 18.

68. Akera, *Calculating a Natural World*, 315; Norberg and O'Neill, *Transforming Computer Technology*

69. Fernando J. Corbató, interview by Arthur L. Norberg, November 14, 1990, Cambridge, Massachusetts. CBI, OH 162, 7. http://purl.umn.edu/107230.

70. Fernando Corbato, Marjorie Merwin-Daggett, and Robert C. Daley, "An Experimental Time-Sharing System," *AIEE-IRE '62 (Spring) Proceedings of the May 1–3, 1962, Spring Joint Computer Conference* (1962).

71. Alan Perlis, "The American Side of the Development of Algol," *ACM SIGPLAN Notices* 13, no. 8 (1978): 4.

72. Ibid.

73. Ensmenger, *The Computer Boys Take Over*; Paul Ceruzzi, "Electronics Technology and Computer Science, 1940–1975: A Coevolution," *IEEE Annals of the History of Computing* 10, no. 4 (1988).

74. Allen Newell, A. J. Perlis, and Herbert Simon, "Computer Science," *Science* 157, no. 3795 (1967).

75. On J. C. R. Licklider, see Norberg and O'Neill, *Transforming Computer Technology*; M. Mitchell Waldrop, *The Dream Machine: J.C.R. Licklider and the Revolution that Made Computing Personal* (New York: Viking Penguin, 2001).

76. Licklider, interview by William Aspray and Arthur Norberg, October 28, 1988, Cambridge, MA. CBI OH 150, 13. http://purl.umn.edu/107436. On cybernetics, see: Steven Heims, *The Cybernetics Group* (Cambridge, MA: The MIT Press, 1991); Peter

Galison, "The Ontology of the Enemy: Norbert Wiener and the Cybernetic Vision," *Critical Inquiry* 21, no. 1 (1994).

77. Licklider, interview by William Aspray and Arthur Norberg, October 28, 1988, Cambridge, MA. CBI OH 150, 13. http://purl.umn.edu/107436.

78. Wesley Clark, interview by Judy O'Neill, May 3, 1990, New York, NY. CBI OH 195. http://purl.umn.edu/107217. Licklider, interview by William Aspray and Arthur Norberg, October 28, 1988, Cambridge, MA. CBI OH 150, 13. http://purl .umn.edu/107436.

79. J. C. R. Licklider, "Man-Computer Partnership," *International Science and Technology* 41(1965): 18.

80. On the rivalry between these two approaches to cybernetics see J. Guice, "Controversy and the State," *Social Studies of Science* 28, no. 1 (1998).

81. Quoted in Chigusa Ishikawa Kita, "JCR Licklider's Vision for the IPTO," *IEEE Annals of the History of Computing*, no. 3 (2003): 68.

82. J. C. R. Licklider, "Man-Computer Symbiosis," *IRE Transactions on Human Factors in Electronics* HFE-1(1960): 4.

83. Ibid, 5.

84. Ibid., 7.

85. Quoted in Waldrop, *The Dream Machine*, 179.

86. Oettinger, Teager, and Griffith, "Report of the Ad Hoc Study Panel on Non-Numerical Info Processing," 11 December 1962, NARA, RG 350, Box 130, Folder "Computer."

87. Ibid.

88. Ibid.

89. Ibid.

90. Minsky to Wiesner, 15 March 1962, "Computer" Box 130, RG 359, NARA.

91. Minsky to Wiesner, 15 March 1962, NARA RG 359, Box 130, Folder "Computer."

92. Ibid.

93. McCarthy to Wiesner, 5 February 1962, NARA, RG 359, Box 130, Folder "Computer."

94. McIlroy to Pierce, 11 January 1962, NARA RG 359, Box 130, Folder "Computer."

95. Goldstine to Pierce, "Comments on Ad Hoc Report on Non-Numerical Processing (PSAC)" 11 January 1962, NARA RG 359, Box 130, Folder "Computer."

96. Teager, Oettinger, Griffith, "PSAC Report of the Ad Hoc Study Panel on Non-Numerical Information Processing," circa September 1962. NARA, RG 359, Box 130, Folder "Computer."

97. Ibid.

98. Ibid.

99. Ibid.

100. The timing of PSAC's study suggests that proposals for a new institute were likely related to the establishment of IPTO, but I have not found conclusive evidence establishing a relation.

101. Richard Barber, "History of DARPA," (Institute for Defense Analysis, 1975), V-48–V-52; Norberg and O'Neill, *Transforming Computer Technology*, 11–13.

102. York, *Making Weapons, Talking Peace*, 171.

103. Quoted in Barber, "History of DARPA," V-51.

104. Waldrop, *The Dream Machine*, 151.

105. Quoted in Barber, "History of DARPA," V-51.

106. Licklider, interview by William Aspray and Arthur Norberg, October 28, 1988, Cambridge, MA. CBI OH 150, 29. http://purl.umn.edu/107436.

107. Fano, interview by Arthur L. Norberg, April 20, 1989, Cambridge, MA. CBI, OH 165, 13. http://purl.umn.edu/107281.

108. Ibid.

109. Ibid.

110. Norberg and O'Neill, *Transforming Computer Technology*. J. A. N. Lee, "The Project MAC Interviews," *IEEE Annals of the History of Computing* 14, no. 2 (1992): 25.

111. Waldrop, *The Dream Machine*, 240.

112. Licklider, interview by William Aspray and Arthur Norberg, October 28, 1988, Cambridge, MA. CBI OH 150, 13. http://purl.umn.edu/107436.

113. Quouted in Waldrop, *The Dream Machine*, 240.

114. Kita, "JCR Licklider's Vision for the IPTO." Waldrop, *The Dream Machine*; Norberg and O'Neill, *Transforming Computer Technology*, 115. Norberg and O'Neill, *Transforming Computer Technology: Information Processing for the Pentagon, 1962–1986*, 290.

116. *Department of Defense Appropriations for 1964; Part 6, Research, Development, Test, and Evaluation*, 236. Emphasis added.

117. Ruina, interview by Slayton, December 9, 2003, Cambridge, MA.

118. Ruina, interview by Slayton, October 8, 2007, Cambridge, MA.

119. Barber, "History of DARPA," V-19.

120. Ibid.

121. Ibid., V-18.

122. Ibid., V-51. PSAC AICBM Panel, "Report of the AICBM Panel," 21 May 1959. DNSA NH01357.

123. Ibid., V-22.

124. ODDR&E [Office of the Director of Defense Research & Engineering] Assessment of Ballistic Missile Defense Program, 17 April 1961, DNSA NH1388.

125. Ibid.

126. See, for example, John Finney, "Pentagon Urges Atom Tests in Air for New Arsenal," *New York Times*, November 2, 1961.

127. McNamara to Kennedy, "Program for Deployment of Nike Zeus," 30 September 1961, DNSA NH01391.

128. Ibid.

129. Panel to Wiesner, "Limited Deployment, Nike-Zeus," 21 October 1961, DNSA NH01392.

130. Ruina interview by Slayton, December 9, 2003, Cambridge, MA.

131. Ibid.

132. Ibid.

133. Ibid.

134. Jack Raymond, "Pentagon to Ask 51 Billion Budget for Coming Year," *New York Times*, December 6, 1961; Jack Raymond, "Nike Zeus Center of Defense Clash," *New York Times*, December 17, 1961.

135. Theodore Shabad, "Russian Reports Solving Rocket Defense Problem," *New York Times*, October 24, 1961.

136. Editors, "Nike-Zeus 'Downs' a Missile in Flight for the First Time," *New York Times*, 22 December 1961. Tests are described in Bell Laboratories, "ABM Research and Development at Bell Laboratories; Project History (Volume I of III)" (Whippany, NJ: Bell Labs, on Behalf of Western Electric, for U.S. Army Ballistic Missile Defense Systems Command, 1975).

137. Theodore Shabad, "Khrushchev Says Missile Can 'Hit a Fly' in Space," *New York Times*, July 17, 1962; John W. Finney, "Nike Zeus Intercepts a Missile Fired From U.S. Over Pacific," *New York Times*, July 20, 1962.

138. Edward Randolph Jayne, "The ABM Debate: Strategic Defense and National Security" (PhD diss., Massachusetts Institute of Technology, 1969), 172.

139. Dustin, "Comments on Some AICBM Issues of the Day," 4 October 1962, DNSA NH01399.

140. Lawrence Legere to Mr. Bundy, "Secretary McNamara's Memorandum for the President Entitled Ballistic Missile Defense," 26 November 1962, DNSA NH01400.

141. John D. Morris, "Senators Meet in Secret, then Bar Nike Speed-Up," *New York Times*, April 12, 1963.

142. Barber, "History of DARPA," V-30.

143. Quoted in ibid., V-18.

144. Gladwin Hill, "For Want of Hyphen, Venus Rocket Is Lost," *New York Times*, July 28, 1962.

Chapter 4

1. Emanuel Adler, "The Emergence of Cooperation," *International Organization* 42, no. 1 (1992); Matthew Evangelista, *Unarmed Forces: The Transnational Movement to End the Cold War* (Cornell University Press, 1999).

2. For realism, see Steve Weber, "Realism, Detente, and Nuclear Weapons," *International Organization* 44, no. 1 (1990). For constructivism, see Adler, "The Emergence of Cooperation."

3. For contemporary coverage, see "Nike-Zeus 'Downs' a Missile in Flight for the First Time," *New York Times*, December 22, 1961; Theodore Shabad, "Russian Reports Solving Rocket Defense Problem," *New York Times*, October 24, 1961; John W. Finney, "Nike Zeus Intercepts a Missile Fired From U.S. Over Pacific," *New York Times*, July 20, 1962; John W. Finney, "Hydrogen Blast Fired 200 Miles Above the Pacific," *New York Times*, July 10, 1962; Jack Raymond, "Nike Intercepts Atlas in flight," *New York Times*, December 13, 1962.

4. Glenn Seaborg, *Kennedy, Khrushchev and the Test Ban* (Berkeley: University of California Press, 1981).

5. Senate Foreign Relations Committee, *Nuclear Test Ban Treaty*, 88th Cong., 1rst sess. August 12–27, 1963, 615.

6. Ibid., 424.

7. Ibid., 570.

8. Ibid., 758–759, 854.

9. Ibid., 759.

10. Ibid.

11. Ibid., 762.

12. Ibid.

13. Abelson to York, 16 September 1963. York Papers, Box 6, Folder 1.

14. York to Abelson, 25 September 1963, York Papers, Box 6, Folder 1.

15. Quoted in Spencer Weart, *Nuclear Fear* (Cambridge, MA: Harvard University Press, 1988), 278.

16. For more on Piel, see Gerard Piel, *The Age of Science: What Scientists Learned in the 20th Century* (New York: Basic Books, 2001), xix.

17. York, Interview by author, La Jolla, CA, August 4, 2006.

18. Jerome Wiesner and Herbert York, "National Security and the Nuclear-Test Ban," *Scientific American* 211, no. 4 (1964): 27–28.

19. Ibid., 32–33.

20. Ibid., 33.

21. Ibid., 35.

22. Ibid.

23. York, interview by author, La Jolla, CA, August 4, 2006.

24. Howard Simons, "Arms Race Called Road to Oblivion," *Washington Post*, September 24, 1964; "Spiral to Oblivion," *New York Times*, September 26, 1964.

25. Piel to Wiesner, 11 November 1964, MC420, Box 100, Folder Scientific American.

26. Nitze to York, 15 December 1964, MC420, Box 100, Folder Scientific American. See also John E Ullman, "Letters," *Scientific American* 212, no. 2 (1965). Some scientists objected to an erroneous chart in the article; York acknowledged the point, but it made no impact on the argument.

27. Ibid.

28. Ruina, interview with author, Cambridge, MA, December 9, 2003.

29. Ruina, interview with author, Cambridge, MA, December 9, 2003.

30. Ruina, interview with author, Cambridge, MA, October 8, 2007.

31. Jack Ruina and Murray Gell-Mann, "Ballistic Missile Defence and the Arms Race." Paper presented at the Proceedings of the Twelfth Pugwash Conference on Science and World Affairs, Udaipur, India, January 27–February 1, 1964), 234.

32. Ibid., 234–235.

33. Ruina, interview with author, Cambridge, MA, October 8, 2007.

34. Robert Kleiman, "3-Year Moratorium Urged on Antimissile Missiles," *New York Times*, November 24, 1965.

35. Both the State Department and Arms Control and Disarmament Agency expressed opposition. John Finney, "Peace Plan Splits Parley in Capital," *New York Times*, December 1, 1965.

36. "Damage Limiting: A Rationale for the Allocation of Resources by the US and the USSR," 21 January, 1964, DNSA, NH00165. See also Edward Randolph Jayne, "The ABM Debate: Strategic Defense and National Security" (PhD diss., Massachusetts Institute of Technology, 1969), 226–227; Fred Kaplan, *The Wizards of Armageddon* (New York: Simon & Schuster, 1983). 320–322.

37. Jayne, "The ABM Debate," 254–256. Richard Barber, "History of DARPA" (Institute for Defense Analysis, 1975), VI-20.

38. See Jayne, "The ABM Debate," 260–261. Fears were hardly alleviated when the Soviets paraded interceptors through Moscow. Henry Tanner, "Soviet Parades Six New Rockets at Celebration," *New York Times*, November 8, 1964.

39. PSAC Strategic Military Panel, "Report on the Proposed Army-BTL Ballistic Missile Defense System," October 29, 1965, DNSA, NH01411.

40. See Ann Finkbeiner, *The Jasons* (New York: Viking, 2006). For a list of PSAC members and terms, see Zuoyue Wang, *In Sputnik's shadow: The President's Science Advisory Committee and Cold War America* (New Brunswick, NJ: Rutgers University Press, 2008).

41. Goldberger, interview by Finn Aaserud, Pasadena, CA, February 12, 1986. Published online by the American Institute of Physics: http://www.aip.org/history/ohilist/4630.html,

42. Goldberger, "Draft of Panel Deliberations on Light Attack Defense Systems," Strategic Military Panel, PSAC, 30 August 1965, NSA, Nuclear History/Berlin Crisis Collection, Box 35, Folder RG 359, 21–22. Hereafter, "Goldberger Draft."

43. Goldberger Draft, 6.

44. Goldberger Draft 6–7.

45. Goldberger Draft, 23–27.

46. Goldberger Draft, 28.

47. Goldberger Draft, 28.

48. Goldberger to Keeny, 30 August 1965, NSA, Nuclear History/Berlin Crisis Collection, Box 35, RG 359.

49. Strategic Military Panel, PSAC, "Report on the Proposed Army-BTL Ballistic Missile Defense System," 29 October 1965, DNSA NH01411.

50. Ibid.

51. "Record of Meeting on DOD FY 1967 Budget (November 9, 1965)," 10 November 1965, DNSA NH00461.

52. See discussion in Anne Hessing Cahn, "American Scientists and the ABM: A Case Study in Controversy," in *Scientists and Public Affairs*, ed. Albert H. Teich (Cambridge, MA: MIT Press, 1974), 61–64.

53. "Ballistic Missile Defense," Report of the Defense Science Board Task Force, 15 September 1966, with cover letter Frederick Seitz (Chairman of the DSB) to the Secretary of Defense, 8 February 1967, DNSA NH01413. Hereafter "DSB Report."

54. DSB Report, 3.

55. DSB Report.

56. DSB Report, v–vi.

57. DSB Report, iv.

58. DSB Report, 9.

59. See Kaplan, *The Wizards of Armageddon*, 320.

60. DSB Report, 7.

61. DSB Report, 5.

62. Senate Committee on Armed Services and Committee on Appropriations, *Military Authorizations and Defense Appropriations for Fiscal Year 1967*, 89th Cong., 2nd sess., February 23, 25, 28, March 8–10, 24, 25, 29–31, 1966, 454.

63. House Committee on Appropriations, *Department of Defense Appropriations for FY 1967, Part 5*, 89th Cong., 2nd sess. April 5, 6, 18–21, 26, 27, 1966, 38.

64. John Norris, "Fight for Curbs on McNamara Opens in House," *Washington Post*, May 18, 1966. See also William White, "The Real Issue: Should Congress Run the Pentagon?," *Washington Post*, April 28, 1966. See also Morton Halperin, "The Decision to Deploy the ABM: Bureaucratic and Domestic Politics in the Johnson Administration," *World Politics* 25, no. 1 (1972).

65. Cyrus Vance to President Johnson, 10 December 1966, DNSA NH01417.

66. Jayne, "The ABM Debate: Strategic Defense and National Security," 338–343; Robert Kleiman, "To Turn Down the Buildup," *New York Times*, September 8, 1968. Robert McNamara, *Blundering Into Disaster* (New York: Pantheon Books, 1986), 55–57.

67. Herbert York, *Race to Oblivion: a Participant's View of the Arms Race* (New York: Simon and Schuster, 1970), 194–195.

68. Bernard Gwerzman, "U.S. Aides in a Quandary," *New York Times*, February 5, 1967; William Beecher, "Missile Defense in Space Studied," *New York Times*, May 31, 1967; UPI, "Text of McNamara Speech on Anti-China Missile Defense and U.S. Nuclear Strategy," *New York Times*, September 19, 1967; Jayne, "The ABM Debate: Strategic Defense and National Security," 375.

69. Anatoly Dobrynin, *In Confidence: Moscow's Ambassador to America's Six Cold War Presidents (1962–1986)*, 1st ed. (New York: Random House, 1995). 166; Glenn Seaborg, *Stemming the Tide: Arms Control in the Johnson Years* (Lexington, MA: Lexington Books, 1987), 426–428.

70. "Transcript of the News Conference Held by Premier Kosygin at the United Nations," *New York Times*, June 26, 1967.

71. See Jayne, "The ABM Debate: Strategic Defense and National Security," 372–395. Kaplan, *The Wizards of Armageddon*, 347; Halperin, "The Decision to Deploy the ABM: Bureaucratic and Domestic Politics in the Johnson Administration."

72. Walt Rostrow to the President, 2 August 1967. DNSA NH01427.

73. "Announcement of ABM Deployment Decision," 5 September 1967, DNSA NH01430. McNamara also sought to ease the anxieties of Japan and European allies.

74. UPI, "Text of McNamara Speech on Anti-China Missile Defense and U.S. Nuclear Strategy."

75. James Reston, "Washington: The Anti-Republican Missile," *New York Times*, September 22, 1967.

76. William Beecher, "U.S. Will Deploy Missile Defense Around Nation: Long Debate Ends," *New York Times*, September 16, 1967; Tom Wicker, "In the Nation: A Nightmare Debate," *New York Times*, September 19, 1967; "The Wrong Race," *Washington Post*, September 19, 1967.

77. Why McNamara left office is still a matter of some contention. In the film *Fog of War*, he said, "Even to this day . . . I don't know whether I quit or was fired?" See transcript at:http://www.errolmorris.com/film/fow_transcript.html.

78. The *New York Times* subtitled its front-page article: "Long Debate Ends."

79. See discussion in Anne Cahn, "Eggheads and Warheads" (PhD diss., Massachusetts Institute of Technology, 1971).

80. Ibid., 90–91.

81. Hans Bethe and Richard Garwin, "Antiballistic-Missile Systems," *Scientific American*, March 1968.

82. Ibid., 23.

83. Ibid., 21.

84. Ibid., 31.

85. See, for example, "Experts See 'Thin' ABM Vulnerable," *Washington Post*, March 3, 1968; Richard Garwin and Hans Bethe, "U.S. Antimissile System Only Gives Illusion of Protection," *Los Angeles Times*, March 3, 1968; Lee Edson, "Scientific Man for All Seasons," *New York Times*, March 10, 1968.

86. William Beecher, "Pentagon Is Studying Improvements in Sentinel," *New York Times*, March 27, 1968.

87. Harriet L. Phelps, "News Items—Garwin-Bethe article," *Public Interest Report*, April 1968.

88. Finkbeiner, *The Jasons*, 90.

89. D. S. Greenberg, "Kistiakowsky Cuts Defense Department Ties over Vietnam," *Science* 159, no. 3818 (1968).

90. Elinor Langer, "After the Pentagon Papers," *Science*, November 26, 1971, 927.

91. Jonathan Allen, *March 4th: Scientists, Students, and Society* (Cambridge, MA: MIT Press, 1970).

92. Quoted in ibid., 142.

93. Quoted in ibid.

94. Quoted in ibid., 149–150.

95. "Scientists Praise Government Ties," *Stanford Daily* 1969. York to Edwin Lennox, 17 March 1969, York Papers, Box 44, Folder 8.

96. "Nuclear Missiles near Cities? Statement of Chicago Chapter of FAS, Nov. 30 1968," FAS Newsletter, November 1968, vol. 21, no. 9, 1–3.

97. See, for example, "A-Missile Sites in Western Suburbs," *Chicago Tribune* November 15, 1968, A1; "Scientist Fears Nuclear Base Here Could Draw Attack," *Chicago Tribune*, November 16, 1968, 3.

98. FAS Newsletter, December 1968. See also Harriet L. Phelps, "The National FAS—Whither and Why?," *Public Interest Report*, February–March 1969. See discussion in Cahn, "American Scientists and the ABM: A Case Study in Controversy," 53–55; Joel Primack and Frank von Hippel, *Advice and Dissent: Scientists in the Political Arena* (New York: Basic Books, 1974), 180.

99. Joseph Boyce, "US Battles Foes of Atom Sites in Area," *Chicago Tribune*, November 25, 1968, A2.

100. Elizabeth Drew, "Reports: Washington," *The Atlantic*, vol. 224, December 1969, 4–18. Quoted in Cahn, "American Scientists and the ABM: A Case Study in Controversy," 57.

101. John W. Finney, "Senate Rejects Missiles Delay," *New York Times*, October 3, 1968.

102. *Department of Defense Appropriations, 1969*, 90th Cong. 2nd sess. *Congressional Record* 114 (October 2, 1968): 29172–29179

103. Bryce Nelson, "ABM: Senators Request Outside Scientific Advice in Closed Session," *Science* 162, no. 3860 (1968).

104. Cahn, "Eggheads and Warheads," 85.

105. John Finney, "Nixon Aide Denies Sentinel Imperils Atom Talks," *New York Times*, March 7, 1969; Senate Foreign Relations Committee, *Strategic and Foreign Policy Implications of ABM Systems, Part I*, 91rst Cong. 1rst sess., March 6, 11, 13, 21, 26, and 28 1969, 6. See also Smith to the Secretary of State, "ABM Decision-Action Memorandum," 6 March 1969, DNSA PR00297.

106. *Strategic and Foreign Policy Implications of ABM Systems, Part I*, 39.

107. Ibid., 35.

108. Ibid., 77.

109. Ibid., 78.

110. Ibid.

111. Ibid., 95.

112. Ibid.

113. Ibid., 96.

114. Ibid.

115. Ibid.

116. Ibid., 97.

117. Ibid.

118. Many critics cited the conclusions of a Pentagon study demonstrating that military electronics systems were falling short of promised performance and yet cost far more than initially expected. Bernard Nossiter, "Weapons Systems: A Story of Failure," *Washington Post*, January 26, 1969.

119. John Finney, "Nixon is Delaying Missile Decision; Opposition Rises," *New York Times*, March 12, 1969. See also William Chapman, "3 Noted Scientists Oppose ABM as Likely to Decrease Security," *Washington Post*, March 12, 1969.

120. Quoted in the *Baltimore Sun*, March 13, 1969.

121. York to Starbird, 21 March 1969, York Papers, Box 44, Folder 7.

122. Starbird to York, 29 March 1969, York Papers, Box 44, Folder 7.

123. Starbird to York, 29 March 1969, York Papers, Box 44, Folder 7.

124. Enthoven to Laird, "Deployment of Sentinel," DNSA PR00363. See also Enthoven to Nitze, "Sentinel Deployment," 14 May 1969, DNSA NH0437.

125. Evangelista, *Unarmed Forces: The Transnational Movement to End the Cold War*, 209–211.

126. Andrei D. Sakharov, "Text of Essay by Russian Nuclear Physicist Urging Soviet-American Cooperation," *New York Times*, July 22, 1968; Theodore Shabad, "A Russian Physicist's Plan: U.S.-Soviet Collaboration," *New York Times*, July 22, 1968.

127. Peter Groses, "U.S. and Soviet Agree to Parleys on Limitation of Missile Systems," *New York Times*, July 2, 1968; Seymour Topping, "Extended Talks on Missiles Seen," *New York Times*, July 27, 1968.

128. See discussion in Gerard Smith, *Doubletalk: The Story of the First Strategic Arms Limitation Talks* (New York: Doubleday Books, 1980).

129. William Beecher, "Laird Supports Antimissile Net," *New York Times*, January 31, 1969. Nixon had recently suggested that "sufficiency" might be a better term than either "superiority or parity," and Laird agreed.

130. William Beecher, "Sentinel Project halted by Laird pending review," *New York Times*, February 7, 1969.

131. John Finney, "Foes of Sentinel in Senate Claim Majority Against It," *New York Times*, February 25, 1969.

132. Robert B. Semple, "Nixon for Limited Missile Plan to Protect U.S. Nuclear Bases," *New York Times*, March 15, 1969.

133. Kissinger to Nixon, "Modified Sentinel System," March 5, 1969 (document in the NSA's Electronic Briefing Book). http://www.gwu.edu/~nsarchiv/NSAEBB/NSAEBB36/18-01.htm. See also conversation, 5 March 1969, DNSA KT00008.

Chapter 5

1. Licklider to Wiesner, "Interim Report on Problems in the Development of Very Large and Complex Systems," 7 March 1969, MC420, Box 19, Folder 11.

2. Licklider to Wiesner, "Underestimates and Overexpectations in the Development of Complex Systems," 24 March 1969, MC420, Box 19, Folder 11.

3. Ibid.

4. Michael Mahoney, "Finding a History for Software Engineering," *IEEE Annals of the History of Computing* (2004); Nathan J. Ensmenger, *The Computer Boys Take Over*

(Cambridge, MA: MIT Press, 2010); Ulf Hashagen, Reinhard Keil-Slawik, and Arthur L Norberg, eds., *History of Computing: Software Issues* (Paderborn, Germany: Springer, 2000); Thomas Haigh, "Dijkstra's Crisis: The End of Algol and Beginning of Software Engineering, 1968-72," in *Software for Europe Project Meeting* (Leiden, Holland, 2010).

5. Atsushi Akera, *Calculating a Natural World: Scientists, Engineers, and Computers During the Rise of U.S. Cold War Research,* Inside Technology series (Cambridge, MA: MIT Press, 2006).

6. Fernando Corbato, Marjorie Merwin-Daggett, and Robert C. Daley, "An Experimental Time-Sharing System," *AIEE-IRE '62 (Spring) Proceedings of the May 1–3, 1962, Spring Joint Computer Conference* (New York: ACM, 1962).

7. Arthur L. Norberg and Judy E. O'Neill, *Transforming Computer Technology: Information Processing for the Pentagon, 1962–1986,* ed. Merritt Roe Smith, Johns Hopkins Studies in the History of Technology (Baltimore: Johns Hopkins University Press, 1996).

8. Ibid.; Judy O'Neill, "'Prestige Luster' and 'Snow-Balling Effects': IBM's Development of Computer Time-Sharing," *IEEE Annals of the History of Computing* 17, no. 2 (1995).

9. Licklider to Smullin, December 19, 1968, quoted in Norberg and O'Neill, *Transforming Computer Technology: Information Processing for the Pentagon, 1962–1986*: 109, n 64.

10. Licklider to Wiesner, "Planning for Multics," December 27, 1968, quoted in ibid., 109–110, n67.

11. Quoted in Akera, *Calculating a Natural World: Scientists, Engineers, and Computers During the Rise of U.S. Cold War Research,* 333–334.

12. Quoted in Norberg and O'Neill, *Transforming Computer Technology: Information Processing for the Pentagon, 1962–1986,* 109.

13. Ibid., 110.

14. Dennis M. Ritchie, "The Evolution of the UNIX Time-Sharing System," *AT&T Bell Laboratories Technical Journal* 63, no. 6 (1984). See also D. M. Ritchie and K. Thompson, "The UNIX Time-Sharing System," *Communications of the ACM* 17, no. 7 (1974).

15. F. J. Corbato, C. T. Clingen, and J. H. Saltzer, "Multics—the First Seven Years." Paper presented at the 1972 Proceedings of the Spring Joint Computer Conference, Atlantic City, May 6–8, 1972.

16. Norberg and O'Neill, *Transforming Computer Technology: Information Processing for the Pentagon, 1962–1986*: 116–117. Werner L. Frank, "Software for Terminal-Oriented Systems," *Datamation,* June 1968.

17. Emerson W. Pugh, *Building IBM: Shaping an Industry and Its Technology* (Cambridge, MA: MIT Press, 1995).

18. The Computer History Museum has placed several early advertisements of System 360 online. See http://archive.computerhistory.org/resources/text/IBM/IBM.System _360.1964.102646088.pdf; http://archive.computerhistory.org/resources/text/IBM/ IBM.360.1964.102646246.pdf.

19. Pugh, *Building IBM*, 275.

20. Computer History Museum: http://archive.computerhistory.org/resources/text/ IBM/IBM.System_360.1964.102646088.pdf.

21. Pugh, *Building IBM*, 277.

22. Ibid.

23. Ibid., 295.

24. http://ldworen.net/fun/os360obit.html. Leonard Woren found the obituary floating around the UCLA Computer Club in the early 1970s. Woren to author by e-mail, May 21, 2011.

25. Frederick Brooks, *The Mythical Man-Month: Essays on Software Engineering*, anniversary ed. (Reading, MA: Addison-Wesley, 1995), 14.

26. Ibid., 25.

27. Ibid., 13.

28. Friedrich Bauer, quoted in Donald MacKenzie, *Mechanizing Proof: Computing, Risk, and Trust* (Cambridge, MA: MIT Press, 2001), 34. Rabi sought to put computers in the service of physicists; see Jean Ford Brennan, *The IBM Watson Laboratory at Columbia University: A History* (New York: International Business Machines Corporation, 1971). Records from the NATO-sponsored study can be found in Rabi's Papers, Box 41, Folder "4-NATO Science Committee Reports, October December."

29. On the origins of the phrase "software engineering," see Maria Eloina Pelaez Valdez, "A Gift from Pandora's Box: The Software Crisis" (University of Edinburgh, 1988); MacKenzie, *Mechanizing Proof*, 34.

30. Peter Naur and Brian Randell, "Software Engineering: Report on a Conference Sponsored by the NATO Science Committee," (Garmisch, Germany, 1969), 120.

31. Ibid.

32. Ibid.

33. Ibid., 121.

34. Ibid.

35. Ibid.

36. Ibid.

37. Ibid., 122.

38. Ibid., 114.

39. Ibid., 122.

40. Operating System 360 is mentioned on 15 separate pages, more than any other single project. Various time-sharing projects are also discussed on 15 different pages.

41. Naur and Randell, "Software Engineering," 132–133.

42. Ibid., 120.

43. Ibid., 122.

44. Brian Randell, "Software Engineering in 1968," Proceedings of the Fourth International Conference on Software Engineering (1979), 5–6.

45. Missile defense is not mentioned anywhere in the summary of discussions; Randell believes that it would likely have been recorded if discussed. Randell to author, e-mail, August 27, 2006.

46. "Sen. Kennedy Hires Panel on Missiles," Washington Post, February 20, 1969.

47. Undated list, MC420, Box 19 (ABM), Folder 10 (no label).

48. Kendall to Weinberg, 17 March 1969, MC420, Box 19, Folder "ABM January 69–May 69."

49. Ibid.

50. Ibid.

51. Licklider to Wiesner, "Interim Report . . ." 7 March 1969, MC420, Box 19, Folder 11.

52. David to Wiesner, 24 March 1969, MC420, Box 19, Folder 5.

53. Licklider to Wiesner, "Interim Report on Problems in the Development of Very Large and Complex Systems," 7 March 1969, MC420, Box 19, Folder 11.

54. Licklider to Wiesner, "Underestimates and Overexpectations in the Development of Complex Systems," 24 March 1969, MC420, Box 19, Folder 11.

55. Ibid.

56. Ibid.

57. Ibid.

58. Ibid.

59. Ibid.

60. This account is based upon McCracken's files, which have been donated to the Charles Babbage Institute in Minneapolis, MN.

61. McCracken, "ABM Diary," October 1969, McCracken files.

62. McCracken to Weizenbaum, Armer, and Williams, 2 June 1969, McCracken files.

63. McCracken, Draft Statement, 14 June 1969, McCracken files.

64. Bernard A. Galler, "ACM President's Letter: 'I Protest,'" *Communications of the ACM* 12, no. 8 (1969).

65. Ibid.

66. Ibid.

67. McCracken to ACM Council, 6 June 1969, McCracken files.

68. Ibid.

69. Ibid.

70. Galler to McCracken, 22 August 1969, McCracken files.

71. Ibid.

72. Ralston to McCracken, 10 June 1969, McCracken files.

73. Ibid.

74. The three were Robert W. Bemer at General Electric, Bernard Galler, and Alan Perlis. A list of targeted sponsors, likely dating to June 1969, can be found in McCracken's files.

75. The conference was mentioned only briefly, after its report was published. Bernard A. Galler, "ACM President's Letter: NATO and Software Engineering?," *Communications of the ACM* 12, no. 6 (1969).

76. ABM Diary, McCracken files.

77. R. V. Head, "Datamation's Glory Days," *IEEE Annals of the History of Computing* 26, no. 2 (2004). McCracken to Forest, 29 May 1969, McCracken files.

78. McCracken to Berkeley, 6 June 1969, McCracken files.

79. Forest to McCracken, 23 July 1969, McCracken files; "Software Testing Not Yet Possible, Licklider States," *Computerworld*, July 16, 1969. Daniel D. McCracken, "The Computer Aspects of a System for Anti-Ballistic Missiles (Letters, Multi-Access Forum)," *Computers and Automation*, July 1969.

80. McCracken statement for the press conference, 30 June 1969; McCracken, "ABM Diary," Oct 1969, McCracken files.

81. McCracken to Wiezenbaum and Williams, 8 June 1969, McCracken files.

82. Thomas Haigh, "Inventing Information Systems: The Systems Men and the Computer, 1950–1968," *Business History Review* 75 (2001).

83. Ensmenger, *The Computer Boys Take Over.*

84. See, for example, Louis Fein, "The Role of the University in Computers, Data Processing, and Related Fields," *Communications of the ACM* 2, no. 9 (1959): 9. Allen Newell, A. J. Perlis, and Herbert Simon, "Computer Science," *Science* 157, no. 3795 (1967).

85. David Parnas, "On the Preliminary Report of C3S," *Communications of the ACM* 8, no. 9 (1966): 242–243. See A. G. Oettinger, "President's Letter to the ACM Membership," *Communications of the ACM* 9, no. 10 (1966).

86. Data on PhD-granting institutions were compiled from the Computer Research Association: http://www.cra.org/resources/forsythe. For more on the algorithm, see Paul Ceruzzi, "Electronics Technology and Computer Science, 1940–1975: A Coevolution," *IEEE Annals of the History of Computing* 10, no. 4 (1988); Ensmenger, *The Computer Boys Take Over*

87. Nathan J. Ensmenger, "The 'Question of Professionalism' in the Computer Fields," *IEEE Annals of the History of Computing* 23, no. 4 (2001): 70.

88. Paul Peters to McCracken, 30 July 1969, McCracken files.

89. Arthur Buskin to Dan McCracken, 15 September 1969, McCracken files.

90. Similar examples include Kenneth Seidel to McCracken, 28 July 1969; Conrad Inngerich to McCracken, 15 July 1969, McCracken files.

91. Geoffrey Lyford to McCracken, 15 July 1969, McCracken files.

92. Ware to McCracken, 27 June 1969, McCracken files. For more on Ware and security, see chapter 7.

93. McCracken, "ABM Diary," October 1969, McCracken files.

94. Hamming to McCracken, 17 October 1969, McCracken files.

95. McCracken to Hamming, 20 February 1970, McCracken files.

96. Nutt to McCracken, 10 June 1969, McCracken files.

97. McCracken to Nutt, 28 July 1969, McCracken files.

98. Oettinger to McCracken, 24 February 1970, McCracken files.

99. McCracken to Arden, 15 April 1970, McCracken files; see also McCracken in *Public Policy and the Expert* (New York: Council on Religion and International Affairs, 1971).

100. McCracken thanked Hanlon the next day, and sent news clippings to all members of the Senate.

101. Bright to McCracken, 16 June 1969, McCracken files.

102. Weizenbaum to Bright, 19 June 1969, McCracken files.

103. Bright to McCracken and Weizenbaum, 20 July 1969, McCracken files.

104. Weizenbaum to Bright, 28 July 1969, McCracken files.

105. Ibid.

106. Ibid.

107. Bright to Weizenbaum, 30 July 1969, McCracken files.

108. James Adams to McCracken, 24 June 1969, McCracken files.

109. Kurt Fuchel to McCracken, 25 June 1969, McCracken files.

110. Salton to McCracken, 25 June 1969, McCracken files.

111. Robert Miller to McCracken, 22 June 1969, McCracken files.

112. Head to McCracken, 19 June 1969, McCracken files.

113. Nat Rochester to McCracken, 19 June 1969; James Sweeny to McCracken, 29 July 1969, McCracken files.

114. The 111 letters only include individuals that responded to the letters McCracken mailed and published in the summer of 1969.

115. Gore to McCracken, 17 June 1969, McCracken files.

116. Hart to McCracken, 20 June 1969, McCracken files. See also Fulbright to McCracken, 11 Jun 1969, McCracken files.

117. Representative Richard L. Ottinger of New York, "The ABM's Questionable Technology," *Congressional Record*, vol. 115, no. 118 (July 16, 1969): E 6018.

118. "National Committee is Formed to Oppose the ABM," *New York Times*, April 18, 1969; "A New Coalition Will Oppose ABM," *New York Times*, March 30, 1969.

119. National Citizens Committee press release, 30 June 1969, McCracken files.

120. Senate Foreign Relations Committee, *Strategic and Foreign Policy Implications of ABM Systems, Part II*, 91rst Cong. 1rst sess., May 14 and 21, 1969, 494.

121. Senate Foreign Relations Committee, *Strategic and Foreign Policy Implications of ABM Systems, Part I*, 91rst Cong. 1rst sess., March 6, 11, 13, 21, 26, and 28, 1969, 379.

122. Ibid.

123. Ibid.

124. Ibid., 275.

125. Ibid.

126. Ibid.

127. Ibid.

128. Quoted in William Beecher, "Dr. Foster Sees a Lag in Missiles," *New York Times*, May 13, 1969.

129. "The Case Against Safeguard," *New York Times*, May 13, 1969.

130. Jerome Wiesner and Herbert York, "National Security and the Nuclear-Test Ban," *Scientific American* 211, no. 4 (1964): 34–35.

131. Foster to York, May 19, 1969. York Papers, Box 44, Folder 7.

132. Ibid.

133. Ibid.

134. Trevor Pinch, ""Testing—One, Two, Three . . . Testing!": Toward a Sociology of Testing," *Science, Technology, & Human Values* 18, no. 1 (1993); Donald MacKenzie, *Inventing Accuracy: A Historical Sociology of Ballistic Missile Guidance* (Cambridge, MA: MIT Press, 1990).

135. John Foster, "Safeguard—a Forum of Opinion," *Modern Data* 3, no. 1 (1970).

136. Daniel D. McCracken, "Essay Contest," *Computers and Automation*, March 1971.

137. McCracken to Armer, Wiezenbaum, and Williams, 27 January 1971, McCracken files.

138. "Excerpts from Majority and Minority Reports by Senate Committee on the ABM," *New York Times*, July 8, 1969.

139. Ibid.

140. "Excerpts from Majority and Minority Reports by Senate Committee on the ABM."

141. "Excerpts from Closing Debate in Senate on the Antimissile Issue," *New York Times*, August 7, 1969.

142. Warren Weaver, "Nixon Missile Plan Wins in Senate by a 51-50 Vote," *New York Times*, August 7 1969.

143. In 1971, McCracken went on a nationwide speaking tour focusing on missile defense, and continued to seek a public debate. He also worked with the Federation of American Scientists, and gained a hearing from Congressional staffers.

144. Activism that did not reach its intended goals included a "counter-conference" to oppose basing the ACM annual meeting in Chicago (in protest of police brutality at the 1968 Democratic National Convention) and efforts to influence the 1970 ACM elections for national office, by soliciting and publicizing candidates' views on the social responsibility of computer professionals.

Chapter 6

1. William Beecher, "President Seeks Expansion of ABM," *New York Times*, January 30, 1970; "Mr. Nixon on the ABM," *Washington Post*, February 1, 1970.

2. Lynn to Kissinger, "Bell Labs on Safeguard," April 14, 1970, Document 10 in William Burr, ed., "Missile Defense Thirty Years Ago," December 18, 2000 (hereafter, "Missile Defense Thirty Years Ago"). http://www.gwu.edu/~nsarchiv/NSAEBB/NSAEBB36.

3. Ibid.

4. Ibid.

5. Drell to DuBridge, December 23, 1969, Document 6 in "Missile Defense Thirty Years Ago." Significantly the meeting between Lynn and Bell Labs was mediated by Presidential Science Advisor Lee DuBridge, not long after Nixon ignored the recommendations of Drell and his group.

6. Lynn, "Bell Labs on Safeguard."

7. Nixon's handwritten note on a letter from Kissinger, "Contractor Doubts on Safeguard," April 15, 1970, Document 10 in "Missile Defense Thirty Years Ago."

8. Ibid.

9. See, for example, Steve Weber, "Realism, Detente, and Nuclear Weapons," *International Organization* 44, no. 1 (1990). Peter Haas, "Introduction: Epistemic Communities and International Policy Coordination," *International Organization* 42, no. 1 (1992).

10. Gerard Smith, *Doubletalk: The Story of the First Strategic Arms Limitation Talks* (New York: Doubleday, 1980); R. L. Garthoff, *Detente and Confrontation: American-Soviet Relations from Nixon to Reagan* (Washington, DC: Brookings Institution Press, 1994); Paul Nitze, *From Hiroshima to Glasnost* (New York: Grove Weidenfeld, 1989).

11. See William Burr, ed., "The Secret History of the ABM Treaty," November 8, 2001 (hereafter, "Secret History"), http://www.gwu.edu/~nsarchiv/NSAEBB/NSAEBB60, and "Missile Defense Thirty Years Ago."

12. GAO, "Army's Evaluation of Alternative Designs for Providing Computer Capabilities Needed for SAFEGUARD Antiballistic Missile System," (Washington, DC: General Accounting Office, 1971). http://archive.gao.gov/f0102/089726.pdf.

13. Drell to DuBridge, December 23, 1969, Document 6 in "Missile Defense Thirty Years Ago."

14. Ibid.

15. Beecher, "President Seeks Expansion of ABM."; "Mr. Nixon on the ABM." On Nixon's original goals, see a conversation with Kissinger et al., March 5, 1969, DNSA KT00008; Kissinger to Nixon, "Modified Sentinel Deployment," March 5, 1969, in "Missile Defense 30 Years Ago."

16. Senate Foreign Relations Committee, *Strategic and Foreign Policy Implications of ABM Systems, Part I*, 91rst Cong. 1rst sess., March 6, 11, 13, 21, 26, and 28, 1969, 328–333.

17. Committee on Armed Services, United States Senate, *Military Procurement for Fiscal Year 1971*, 91rst Cong. 2nd sess., May 19, 1970, 2350–2355.

18. Ibid, 2308.

19. Ibid, 2355.

20. Daniel Buchonnet, "MIRV: A Brief History of Minuteman and Multiple Reentry Vehicles," Lawrence Livermore Laboratory, February 1976. http://www.gwu.edu/~nsarchiv/nsa/NC/mirv/mirv.html.

21. Ted Greenwood, *Making the MIRV: A Study of Defense Decision-Making* (Cambridge, MA: Ballinger Publishing Co., 1975), 110.

22. Strategic Military Panel, "Proposal for a Comprehensive Freeze on Strategic Weapons," 6 June 1969, DNSA PR00402.

23. See, for example, House Committee on Appropriations, *Safeguard Antiballistic Missile System*, 91rst. Cong., 1rst. sess. May 22, 1969.

24. Herbert F York, "ABM, MIRV, and the Arms Race," *Science*, July 17, 1970, 258.

25. Herbert York, *Race to Oblivion: A Participant's View of the Arms Race* (New York: Simon and Schuster, 1970), 231.

26. Senate Foreign Relations Committee, *Strategic and Foreign Policy Implications of ABM Systems, Part I*, 267–268.

27. York, *Race to Oblivion*, 231.

28. See 5 March 1969, DNSA KT00008.

29. Kissinger to Nixon, May 23, 1969, Document 1 in "Secret History."

30. Quoted in Strobe Talbott, *The Master of the Game: Paul Nitze and the Nuclear Peace* (New York: Alfred A. Knopf, 1988), 114.

31. *Military Procurement for Fiscal Year 1971*, 2230.

32. Emphasis in original, ibid., 2352.

33. Senate Foreign Relations Committee, *Strategic and Foreign Policy Implications of ABM Systems, Part II*, 91rst Cong. 1rst sess., May 14th and 21 1969, 602.

34. Albert Wohlstetter, "Letters to the Editor," *New York Times*, June 15, 1969.

35. George Rathjens, "Letters to the Editor," *New York Times*, June 22, 1969.

36. William Beecher, "Report on Safeguard ABM Testimony Finds Unprofessional and Misleading Comments on Both Sides," *New York Times*, October 1, 1971.

37. Ibid.

38. Ibid.

39. Ibid.

40. McGeorge Bundy, "Existential Deterrence and Its Consequences," in *The Security Gamble: Deterrence Dilemmas in the Nuclear Age*, ed. Douglas MacLean (Totowa, NJ: Rowman and Allanheld, 1984).

41. McGeorge Bundy, "To Cap the Volcano," *Foreign Affairs* 47 (1969).

42. Ibid.

43. Smith, *Doubletalk*, 29.

44. Nitze, *From Hiroshima to Glasnost*, 295–298.

45. Garthoff, *Detente and Confrontation*; Nitze, *From Hiroshima to Glasnost*; Smith, *Doubletalk*.

46. William Beecher, "Expansion of ABM to 3d Missile Site Is Sought," *New York Times*, February 25, 1970.

47. Helmut Sonnenfeldt to Kissinger, 29 April 1970, Document 4 in "Secret History."

48. John Finney, "Conferees Vote ABM Limitation," *New York Times*, September 25, 1970.

49. Hedrick Smith, "Soviet Is Believed to Favor Limitation on Deploying ABM," *New York Times*, November 29, 1970.

50. K. Wayne Smith and Helmut Sonnenfeldt, to Kissinger, January 12, 1971, Document 6 in "Secret History."

51. Bell Telephone Laboratories, *Ballistic Missile Defense Advanced Development Program—Advanced Data Processing*, 30 September 1969, 16. Copy from McCracken files.

52. Smith, *Doubletalk*, 169.

53. Buchonnet, *MIRV: A Brief History* (note 21), 17.

54. Max Frankel, "Compromise Set," *New York Times*, May 21, 1971. For more details on the negotiations, see "Secret History."

55. Nitze, *From Hiroshima to Glasnost*, 309–310.

56. Ibid., 315–316.

57. Rebecca Slayton, "From Death Rays to Light Sabers: Making Laser Weapons Surgically Precise," *Technology and Culture* 52, no. 1 (2011).

58. State Department Cable to SALT Delegation, "August 9 Verification Panel Meeting," August 10, 1971. Document 17 in "Secret History."

59. Ibid.

60. State Department Cable to SALT Delegation, "SALT Guidance," August 12, 1971, Document 18, "Secret History."

61. "SALT I Negotiating History," System Planning Corporation, March 1985, 21–22. Copy obtained from the National Security Archive.

62. Garthoff, *Detente and Confrontation*, 269–270; Nitze, *From Hiroshima to Glasnost*, 322–328; Smith, *Doubletalk*, 407–440.

63. John Finney, "Senate Approves Pact," *New York Times*, August 4, 1972.

64. Texts, "'Agreed Interpretations' and Unilateral Statements with Arms Pacts," *New York Times*, June 14, 1972.

65. Richard Barber, "History of DARPA" (Institute for Defense Analysis, 1975), VIII-30.

66. GAO, "The Army Reorganization for the 1970s: An Assessment of the Planning" (Washington, DC: General Accounting Office, 1973).

67. Senate Committee on Armed Services, *FY75 Authorization for Military Procurement, Research and Development, and Active Duty, Selected Reserve and Civilian Personnel Strengths. Part 7*, 93rd Cong., 2nd sess., April 4, 5, 12, 16, 23, 25, 26, May 2, 1974, 3412–3425.

68. Initially this was called "hard-point defense." See *Military Procurement for Fiscal Year 1971*, 2346–2351.

69. Committee on Armed Services, U.S. Senate, *Military Implications of the Treaty on the Limitations of Anti-Ballistic Missile Systems and the Interim Agreement on Limitation of Strategic Offensive Arms*, 92nd Cong., 2nd sess. June 6, 20, 22, 28; July 18, 19, 21, 24, and 25, 1972, 282.

70. GAO, "The SAFEGUARD Ballistic Missile Defense System" (Washington, DC: General Accounting Office, 1973), 6, 9.

71. Congressional staffer, quoted in Finlay Lewis, "ABM: Countdown to Oblivion," *Washington Post*, October 5, 1975.

72. Senate Committee on Armed Services, *FY77 Authorization for Military Procurement, Research and Development, and Active Duty, Selected Reserve and Civilian Personnel Strengths. Part 12: Research and Development*, 94th Cong., 2nd sess., March 23, 30, April 5, 6, 7, 1976, 6684.

73. John Finney, "Safeguard ABM System to Shut Down; $5 Billion Spent in 6 Years Since Debate," *New York Times*, November 25, 1975.

74. William K. Stevens, "Abandonment of Safeguard ABM System Stuns the Town of Langdon, ND," *New York Times*, November 25, 1975.

75. *FY77 Authorization for Military Procurement, Research and Development, and Active Duty, Selected Reserve and Civilian Personnel Strengths. Part 12: Research and Development*, 6680.

76. Ibid.

77. Ibid., 6686.

78. For an excellent discussion of PSAC's views, see Anne Cahn, "Eggheads and Warheads" (PhD diss., Massachusetts Institute of Technology, 1971), 199–200.

79. Panofsky, interview by author, July 12, 2006, Stanford, CA.

80. *Strategic and Foreign Policy Implications of ABM Systems, Part I*, 327.

81. Ibid., 328.

82. Senate Foreign Relations Committee, *ABM, MIRV, SALT, and the Nuclear Arms Race*, 91rst Congress, 2nd Session, 1970, 442.

83. Ibid.

84. Ibid., 522–523.

85. Ibid.

86. Ibid., 534–535.

87. Drell interview.

88. *Strategic and Foreign Policy Implications of ABM Systems, Part I*, 312.

89. York to Keith Glennan, 10 April 1969, York Papers, Box 44, Folder 7.

90. York to Panofsky, 10 April 1969, York Papers, Box 44, Folder 7.

91. Drell to Wiesner, 24 September 1969, MC 420, Box 19, Folder 5.

92. Ibid.

93. Ibid.

94. Ibid.

95. Ibid.

96. Gregg Herken, *Cardinal Choices: Presidential Science Advising from the Atomic Bomb to SDI* (New York: Oxford University Press, 1992), 178.

97. Spencer Rich, "SST Gets Nixon's Go-Ahead," *New York Times*, September 24, 1969.

98. Herken, *Cardinal Choices: Presidential Science Advising from the Atomic Bomb to SDI*, 178.

99. Ibid., 177.

100. Christopher Lydon, "Nixon Aide Scores SST Test Design," *New York Times*, August 29, 1970.

101. Christopher Lydon, "Senate Rejects SST Fund," *New York Times*, December 4, 1970.

102. Elinor Langer, "After the Pentagon Papers," *Science*, November 26, 1971, 928.

103. Ibid.

104. Ibid.

105. D. S. Greenberg, "Nixon's Science Adviser," *Science*, October 23, 1970, 417.

106. John Walsh, "Science Adviser: DuBridge Retires, David Nominated as Successor," *Science*, August 28, 1971, 843.

107. Ibid.

108. Ibid.

109. Ibid., 844.

110. Ibid.

111. Philip Abelson, "Departure of the President's Science Adviser," *Science*, January 19, 1973.

112. Deborah Shapley, "Science in Government: Outline of New Team Emerges," *Science*, February 2, 1973.

113. Rodney Nichols, "Mission-Oriented R&D," *Science*, April 2, 1971, 29.

114. Ibid.

115. Barber, "History of DARPA," VIII-19–25.

116. Ibid.

117. Ibid. Emphasis in original.

118. Kelly Moore, *Disrupting Science: Social Movements, American Scientists, and the Politics of the Military, 1945–1975* (Princeton, NJ: Princeton University Press, 2008).

119. Jay Orear, "A Vote for Mansfield," *Physics Today*, May 1970, 11.

120. Ibid.

121. Jonathan Allen, *March 4th: Scientists, Students, and Society* (Cambridge, MA: MIT Press, 1970); Gary L. Downey, "Reproducing Cultural Identity in Negotiating Nuclear Power: The Union of Concerned Scientists and Emergency Core Cooling," *Social Studies of Science* 18, no. 2 (1988); Dorothy Nelkin, *The University And Military Research: Moral Politics at M.I.T.* (Ithaca, NY: Cornell University Press, 1972).

122. Barry M. Caspar, "Physicists and Public Policy," *Physics Today*, May 1974, 32.

123. "Council Establishes Panel on Public Affairs," *Physics Today*, January 1975.

124. Langer, "After the Pentagon Papers," 103.

125. David Kaiser, *American Physics and the Cold War Bubble* (Chicago: University of Chicago Press, in preparation). Book details available at http://web.mit.edu/dikaiser/www/CWB.html.

126. Kenneth Flamm, *Targeting the Computer: Government Support and International Competition* (Washington, DC: The Brookings Institution, 1987). 88–89.

Chapter 7

1. Anthony I. Wasserman, "Letter from the Chairman," *SEN* 2, no. 3 (1977).

2. Michael Mahoney, "Finding a History for Software Engineering," *IEEE Annals of the History of Computing* (2004); Nathan J. Ensmenger, *The Computer Boys Take Over* (Cambridge, MA: MIT Press, 2010), 195–221; Donald MacKenzie, *Mechanizing Proof: Computing, Risk, and Trust* (Cambridge, MA: MIT Press, 2001), 34–41.

3. Thomas Haigh has also noted a disjuncture between the conferences of the late 1960s, and the 1970s discourse of "software engineering." Thomas Haigh, "Dijkstra's Crisis: The End of Algol and Beginning of Software Engineering, 1968-72," Draft Paper for Discussion at the Software-EU Project Meeting in Leiden, Netherlands, 2010. http://tomandmaria.com/tom/Writing/DijkstrasCrisis_LeidenDRAFT.pdf.

4. Martin Campbell-Kelly, *From Airline Reservations to Sonic the Hedgehog* (Cambridge, MA: MIT Press, 2003), 109–115.

5. E. W. Dijkstra, "The Threats to Computing Science." Paper presented at the ACM South Central Regional Conference, Austin, Texas, November 16–18. 1984.

6. E. W. Dijkstra, "What Led to 'Notes on Structured Programming," EWD 1308, June 10, 2001. http://www.cs.utexas.edu/users/EWD/transcriptions/EWD13xx/EWD1308.html.

7. Edsger W. Dijkstra, "Letters to the Editor: Go to Statement Considered Harmful," *Communications of the ACM* 11, no. 3 (1968).

8. See Paul Abrahams, "'Structured Programming' Considered Harmful," *ACM SIG-PLAN Notes* 10, no. 4 (1975).

9. See John R. Rice and Edsger W. Dijkstra, "Letters to the Editor: The Go To Statement Reconsidered," *Communications of the ACM* 11, no. 8 (1968).

10. E. W. Dijkstra, "Notes on Structured Programming," August 1969. http://www.cs.utexas.edu/users/EWD/transcriptions/EWD02xx/EWD268.html. See also Leonard H. Weiner, "The Roots of Structured Programming," *SIGCSE Bulletin* 10, no. 1 (1978); MacKenzie, *Mechanizing Proof.*

11. J. L. Buxton and Brian Randell, *Software Engineering Techniques: Report on a Conference Sponsored by the NATO Science Committee in Rome, Italy, 27th to 31st October 1969* (NATO Science Committee, 1970).

12. F. T. Baker, "Chief Programmer Team Management of Production Programming," *IBM System Journal* 11, no. 1 (1972): 57.

13. Ibid.: 56–57; Harlan Mills, "Top Down Programming in Large Systems," in *Debugging Techniques in Large Systems*, ed. Randall Rustin (Englewood Cliffs, NJ: Prentice-Hall, 1970).

14. Buxton and Randell, "Software Engineering Techniques," 51–52.

15. Ibid., 52–53.

16. Ibid., 85.

17. Ibid., 13.

18. Ibid., 13.

19. Ibid., 8.

20. Quoted in MacKenzie, *Mechanizing Proof*, 37.

21. Campbell-Kelly, *From Airline Reservations to Sonic the Hedgehog*, 109–115.

22. Frank, "Software for Terminal-Oriented Systems."

23. Campbell-Kelly, *From Airline Reservations to Sonic the Hedgehog.*

24. Werner L. Frank, "The History of Myth No.1," *Datamation*, May 1983.

25. GAO, "Acquisition and Use of Software Products for Automatic Data Processing Systems in the Federal Government; Report to the Congress," (Washington, DC: General Accounting Office, 1971).

26. Thomas McNaugher, *New Weapons, Old Politics: America's Military Procurement Muddle* (Washington, DC: Brookings, 1989), 65–73. For more on the practices and problems of audit cultures, see Michael Power, *The Audit Society: Rituals of Verification* (New York: Oxford University Press, 1997).

27. See discussion in McNaugher, *New Weapons, Old Politics*, 69.

28. Benjamin F. Schemmer, "DoD's Computer Critics Nearly Unplug Ailing World Wide Military Command and Control System," *Armed Forces Journal* 108 (1970).

29. This account draws heavily upon analyses in David E. Pearson, *The World Wide Military Command and Control System: Evolution and Effectiveness* (Maxwell Air Force Base, AL: Air University Press, 2000).

30. "How Not to Build a C&C System Is Still Unanswered," *Armed Forces Management* 12, no. 10 (1966), 109.

31. Subcommittee of the Committee on Appropriations, U.S. Senate, *Department of Defense Appropriations for Fiscal Year 1973: Part I*, 92nd Cong., 2nd sess., February 24, 25, 28, March 14, 16, 22–24, May 4, 1972, 1056.

32. George Weiss, "Restraining the Data Monster: The Next Step in C3," *Armed Forces Journal* 108 (1971).

33. "Can Vulnerability Menace Command and Control?," *Armed Forces Management* (1969): 43.

34. Joseph Volz, "Revamped World-Wide Computer Net Readied," *Armed Forces Journal* 107 (1970).

35. Phil Hirsch, "GAO Hits Wimmix Hard; FY'72 Funding Prospects Fading Fast," *Datamation* (1971).

36. "WWMCCS Report, One Year Later," *Government Executive* 3 (1971).

37. Pearson, *The World Wide Military Command and Control System*, 125–132.

38. House Committee on Armed Services, 92nd Cong., 1rst Sess., *Review of Department of Defense World Wide Communications*, May 10, 1971, 11.

39. Ibid., 14.

40. Ibid.

41. Ibid., 17.

42. Nixon, National Security Decision Memorandum (NSDM)-242, "Policy for Planning the Employment of Nuclear Weapons," January 17, 1974. http://www.fas.org/irp/offdocs/nsdm-nixon/nsdm_242.pdf.

43. On Nixon's policy, see D. Ball, "Deja Vu: The Return to Counterforce in the Nixon Administration," (California Seminar on Arms Control and Foreign Policy, 1974).

44. "Minutes of the Verification Panel Meeting held August 9, 1973," Jeanne Davis to Kissinger, 15 August 1973, KT00790, DNSA.

45. Ashton B. Carter, John D. Steinbruner, and Charles A. Zraket, eds., *Managing Nuclear Operations*, (Washington D.C.: The Brookings Institution, 1987), 647.

46. Interestingly, these systems suffered from frequent outages, some computer related. See discussion in *Department of Defense Appropriations for Fiscal Year 1974: Part 6 Procurement*, 1687–1689.

47. House Committee on Armed Services, *Review of Department of Defense World Wide Communications*, 17.

48. The Assistant Secretary was renamed the Director of Telecommunications and Command and Control Systems (DTACCS). See discussion in Pearson, *The World Wide Military Command and Control System*, 165.

49. Jeffrey Richelson, "PD-59, NSDD-13 and the Reagan Strategic Modernization Program," *Journal of Strategic Studies* 6 (June 1983): 125–146.

50. Pearson, *The World Wide Military Command and Control System*, 200–206.

51. GAO, "The World Wide Military Command and Control System—Major Changes Needed in its Automated Data Processing Management and Direction, Report to the Congress" (Washington, DC: General Accounting Office, 1979), ii.

52. Edgar Ulsamer, "The Military Decision-Makers' Top Tool," *Air Force Magazine* (1971); Edgar Ulsamer, "Command and Control is of Fundamental Importance," *Air Force Magazine* (1972); Barry Boehm, "Software and its Impact: A Quantitative Assessment," *Datamation*, May (1973); Lee M. Paschall, "USAF Command Control and Communication Priorities," *Signal* 28 (1973).

53. Boehm, "Software and its Impact: A Quantitative Assessment," 48.

54. Ibid.: 54–55.

55. Ibid.

56. Ibid.: 59.

57. Boehm, interview with author, July 3, 2007, USC. Davis Dyer, *TRW: Pioneering Technology and Innovation since 1900* (Boston, MA: Harvard Business School Press, 1998), 306–307.

58. Harvey M. Sapolsky, *The Polaris System Development: Bureaucratic and Programmatic Success in Government* (Cambridge, MA: Harvard University Press, 1972). Essays in Agatha Hughes and Thomas P. Hughes, eds., *Systems, Experts, and Computers: The Systems Approach in Management and Engineering, World War II and After* (Cambridge, MA: MIT Press, 2000).

59. R. D. Williams, "Managing the Development of Reliable Software," in *Proceedings of the International Conference on Reliable Software* (Los Angeles, CA: ACM, 1975), 7.

60. Boehm, interview with author, July 3, 2007, USC.

61. See Barry Boehm et al., *Characteristics of Software Quality*, 4 vols., vol. 1, TRW Series of Software Technology (New York: North-Holland Publishing Co., 1978); Thomas A. Thayer, *Software Reliability: A Study of Large Project Reality*, 4 vols., vol. 2, TRW Series of Software Technology (New York: North-Holland Publishing Co, 1978).

62. http://web.archive.org/web/20060204221004/http://www.dacs.dtic.mil/techs/history/toc.html. Boehm, "Software and Its Impact: A Quantitative Assessment," 49.

63. Raymond Yeh and Barry Boehm, "Foreword and Program Chairman's Message," in *Proceedings of the International Conference on Reliable Software* (Los Angeles, CA: ACM, 1975).

64. Milton Minneman (chairman), "Foreword." Paper presented at the IEEE Symposium Computer Software Reliability, New York City, 1973.

65. Boehm, interview with author, July 3, 2007, USC. Government and other agencies have sometimes joined as sponsors.

66. For a list of conferences and sponsors, see http://www.icse-conferences.org/history.html.

67. The abstracts of all conference papers were surveyed using ACM's Portal system.

68. Peter Neumann, "Letter from the Editor," *SEN* 2, no. 5 (1977).

69. Boehm, interview with author, July 3, 2007, USC.

70. Wasserman, "Letter from the Chairman."

71. Ibid., no. 1: 1.

72. Thomas Steel, "Letter from the Chairman," *SEN* 1, no. 1 (1976): 2.

73. Peter Neumann, "Letter from the Editor," *SEN* 1, no. 1 (1976): 3.

74. Peter Neumann, "Letter from the Editor," *SEN* 3, no. 1 (1978).

75. William A. Whitaker, "ADA—The Project: The DoD High Order Language Working Group," (1996), 178.

76. Embedded computer systems are distinguished from office computers by the acquisition rules. Weapons systems computers are governed by a DoD 5000.1 series of regulations that apply to major defense acquisitions. Computers used for more routine office tasks (or automatic data processing), are governed by the Brooks ADP Act. Ibid., 177.

77. Ibid.

78. Ibid., 208.

79. Ibid., 186.

80. Ibid.

81. Ibid.

82. Ibid., 229.

83. Edsger W. Dijkstra, "On a Language Proposal for the Department of Defence," September 17, 1975; memo in the Dijkstra archive: http://www.cs.utexas.edu/users/EWD/transcriptions/EWD05xx/EWD514.html.

84. Ibid.

85. Barbara Ryder, Mary Lou Soffa, and Margaret Burnett, "The Impact of Software Engineering on Modern Programming Languages," *ACM Transactions on Software Engineering and Methodology* 14, no. 4 (2005).

86. David L. Parnas, "Building Reliable Software in BLOWHARD," *SEN* 2, no. 3 (1977).

87. Ibid.

88. Peter Neumann, "Panel Discussion on the Limits of Language Design for Reliable Software," *SEN* 2, no. 3 (1977): 13.

89. David Parnas, "Designing Software for Ease of Extension and Contraction," *IEEE Transactions on Software Engineering* (1979); David Parnas, "On the Criteria to be Used in Decomposing Systems into Modules," *Communications of the ACM* 15, no. 12 (1972).

90. Ryder, Soffa, and Burnett, "The Impact of Software Engineering on Modern Programming Languages."

91. E. W. Dijkstra, "America's Programming Plight" EWD 750, September 19, 1980. http://www.cs.utexas.edu/users/EWD/transcriptions/EWD07xx/EWD750.html.

92. E. W. Dijkstra, "Trip Report E.W.Dijkstra, Munich–London, September 16 –29, 1979 (EWD 715)," (1979).

93. Dijkstra, "The Threats to Computing Science."

94. See MacKenzie, *Mechanizing Proof: Computing, Risk, and Trust*, 153–162.

95. Willis Ware, "Security Controls for Computer Systems: Report of the Defense Science Board Task Force on Computer Security," (Washington, DC: Office of the Director of Defense Research and Engineering, Washington, DC, 1970), v.

96. James P. Anderson, "Computer Security Technology Planning Study, Vol. 1," (Hanscom Air Force Base, Bedford, MA: Electronic Systems Division of the Air Force Systems Command, 1972), 3–4.

97. James P. Anderson, "Computer Security Technology Planning Study, Vol. 2" (Hanscom Air Force Base, Bedford, MA: Electronic Systems Division of the Air Force Systems Command, 1972).

98. Anderson, "Computer Security Technology Planning Study, Vol. 1," 28.

99. Donald MacKenzie, *Mechanizing Proof*.

100. Harlan Mills, "How to Write Correct Programs and Know It." Paper presented at the Proceedings of the International Conference on Reliable Software, Los Angeles, CA, 1975, 364.

101. Ibid.

102. Richard A. DeMillo, Richard J. Lipton, and Alan J. Perlis, "Social Processes and Proofs of Theorems and Programs." Paper presented at the Proceedings of the Fourth ACM Symposium on Principles of Programming Languages, January 1977, 211.

103. Ibid., 206.

104. Edsger W. Dijkstra, "On a Political Pamphlet from the Middle Ages," *SEN* 3, no. 2 (1978).

105. Roger Van Ghent, "Social Processes and Proofs of Theorems and Programs," *SEN* 3, no 3 (1978).

106. Quoted in MacKenzie, 169.

107. Ibid, 167–168.

108. Robert Glass, "Computing Failure: A Learning Experience," *SEN* 3, no. 5 (1978).

109. Ibid; Robert Glass, *Software Conflict 2.0: The Art and Science of Software Engineering* (Atlanta, GA: Developer Books, 2006).

110. Peter Neumann, "Letter from the Editor," *SEN* 4, no. 2 (1979): 3.

111. Ibid.: 4.

112. Ibid., no. 3: 2.

113. Jim Horning, "A Note on Program Reliability," *ACM SIGSOFT, Software Engineering Notes* (1979): 6.

114. Ibid.: 7.

115. Ibid.: 8.

116. Peter Neumann, "Letter from the Editor," *SEN* 8, no. 5 (1983): 9.

117. Peter G. Neumann, *Computer-Related Risks* (New York: ACM Press, 1995).

118. Campbell-Kelly, *From Airline Reservations to Sonic the Hedgehog*, 201–202.

119. W. M. McKeeman, "The Role of Software Engineering in the Microcomputer Revolution: An Overview," in *Proceedings of the 4th International Conference on Software Engineering* (Munich, Germany: IEEE Press, 1979).

120. See, for example, Peter Neumann, "Letter from the Editor," *SEN* 6, no. 1 (1981).

121. For a brief discussion of the appropriate technology movement, see Thomas P. Hughes, *American Genesis*, (New York: Penguin Books, 1989): 446–448.

122. Charles Perrow, *Normal Accidents: Living with High-Risk Technologies* (1984; repr., Princeton, NJ: Princeton University Press, 1999).

Chapter 8

1. David Parnas, "Software Aspects of Strategic Defense Systems," *Communications of the ACM* 28, no. 12 (1985): 1328.

2. Herbert York, *Race to Oblivion: a Participant's View of the Arms Race* (New York: Simon and Schuster, 1970, 182.

3. For an excellent account of MX, see John Edwards, *Superweapon* (New York: W.W. Norton & Company, 1982).

4. Hans Bethe and Kurt Gottfreid, "Assessing Reagan's Doomsday Scenario," *New York Times*, April 11, 1982.

5. Charles Mohr, "A Scary Debate Over Launch Under Attack," *New York Times*, July 18, 1982.

6. John F. Burns, "Soviet Starts Making Some Admissions," *New York Times*, September 24, 1983.

7. For definitions of these different policies see Richard Garwin, "Launch Under Attack to Redress Minuteman Vulnerability?," *International Security* 4, no. 3 (1979/80).

8. Richard Halloran, "Reagan Arms Policy Said to Rely Heavily on Communications," *New York Times*, October 12, 1981.

9. Quoted in David E Pearson, *The World Wide Military Command and Control System: Evolution and Effectiveness* (Maxwell Air Force Base, Alabama: Air University Press, 2000), 264.

10. See D. Ball, "US Strategic Forces: How Would They Be Used?," *International Security* 7, Winter (1982); B. G. Blair, *Strategic Command and Control: Redefining the Nuclear Threat* (Washington, DC: Brookings Institution Press, 1985).

11. Gary Hart and Barry Goldwater, "Recent False Alerts from the Nation's Missile Attack Warning System," (Washington, DC: United States Senate, 1980). On how the commands swept some mistakes under the rug, see Scott Sagan, *The Limits of Safety: Organizations, Accidents, and Nuclear Weapons* (Princeton, NJ: Princeton University Press, 1993). 238.

12. Peter Neumann, "Letter from the Editor," *SEN* 5, no. 3 (1980): 4.

13. Paul Bracken, *The Command and Control of Nuclear Forces* (New Haven, CT: Yale University Press, 1983), 220.

14. Richard Halloran, "U.S. Plans Weapons Against Satellites," *New York Times,* June 6, 1982; AP, "Gains by Soviet Reported in Test to Kill Satellites," *New York Times,* March 19, 1981.

15. Committee on Foreign Relations, *Controlling Space Weapons,* 98th Cong., 1rst sess., April 14, May 18, 1983. Charles Mohr, "U.S. Urged to Seek Ban on Weapons in Space," *New York Times,* May 19, 1983.

16. For an analysis of the relation between Reagan's policies and his predecessors, see Jeffrey Richelson, "PD-59, NSDD-13 and the Reagan Strategic Modernization Program," *Journal of Strategic Studies* 6, no. 2 (June 1983): 125–146.

17. Richard Halloran, "Pentagon Draws Up First Strategy for Fighting a Long Nuclear War," *New York Times,* May 30, 1982; Richard Halloran, "Weinberger Defends His Plans on a Protracted Nuclear War," *New York Times,* August 10, 1982.

18. Robert Scheer, *With Enough Shovels: Reagan, Bush, and Nuclear War* (New York: Random House, 1982), 18.

19. For a succinct history of the antinuclear movement, see Frances B. McCrea and Gerald E. Markle, *Minutes to Midnight: Nuclear Weapons Protest in America* (Newbury Park: Sage, 1989).

20. Fox Butterfield, "Anatomy of the Nuclear Protest," *New York Times,* July 11, 1982. On the discursive shift in the 1980s anti-nuclear movement, see Paul Chilton, ed. *Language and the Nuclear Arms Debate: Nukespeak Today* (Dover, NH: Francis Pinter, 1985).

21. CPSR files, Palo Alto, CA.

22. Alan Borning, "Computer System Reliability and Nuclear War," *Communications of the ACM* 30, no. 3 (1987); Richard L. Garwin, "Launch Under Attack to Redress Minuteman Vulnerability?" *International Security* 4, no. 3 (1979/80), 117–139.

23. "Analogy," e-mail from Bill Finzer to "Antiwar" listserve, September 1, 1983, in response to talk by Borning. Files of Rodney Hoffman (CPSR member).

24. Finerman to McCracken, 16 March 1970, McCracken files.

25. See chapter 7.

26. Adele Goldberg, "Reliability of Computer Systems and Risks to the Public," *Communications of the ACM* 28, no. 2 (1985): 131.

27. Ibid.

28. Ibid.

29. John Herbers, "Widespread Vote Urges Nuclear Freeze," *New York Times*, November 4, 1982.

30. McCrea and Markle, *Minutes to Midnight*.

31. George C. Wilson and David Hoffman, "Joint Chiefs had Counseled Regan Against 'Dense Pack,'" *Washington Post*, December 9, 1982.

32. For discussion of these systems, see Senate Committee on Armed Services, *Department of Defense Authorization for Appropriations for FY82. Part 7: Strategic and Theater Nuclear Forces, Civil Defense*, 97th Cong., 1rst Sess., February 18, 20, 25, 27, March 4, 6, 11, 30, 1981.

33. For an excellent summary of these disparate efforts, see Francis Fitzgerald, *Way Out There in the Blue: Reagan, Star Wars, and The end of the Cold War* (New York: Simon & Schuster, 2000).

34. Rebecca Slayton, "From Death Rays to Light Sabers: Making Laser Weapons Surgically Precise," *Technology and Culture* 52, no. 1 (2011).

35. Malcolm Wallop, "Opportunities and Imperatives of Ballistic Missile Defense," *Strategic Review* (1979): 18.

36. Philip M. Boffey, "Pressures Are Increasing for Arms Race in Space," *New York Times*, October 18, 1982.

37. Slayton, "From Death Rays to Light Sabers."

38. For the best documented account of the origins of the "Star Wars" speech, see Hedrick Smith, *The Power Game* (New York: Random House, 1988), 603–616.

39. CNN archive of Reagan's speeches: http://www.cnn.com/SPECIALS/2004/reagan/stories/speech.archive/defense.html.

40. Ibid.

41. Slayton, "From Death Rays to Light Sabers."

42. Richard Garwin, "Reagan's Riskiness," *New York Times*, March 30, 1983.

43. Edward Teller, "Reagan's Courage," *New York Times*, March 30, 1983.

44. Meg Greenfield, "Calling Buck Rogers," *Washington Post*, March 30, 1983.

45. Ibid.

46. Clarence A. Robinson, "Study Urges Exploiting Technologies," *Aviation Week & Space Technology*, October 24, 1983.

47. Senate Foreign Relations Committee, *Strategic Defense and Anti-Satellite Weapons*, 98th Cong., 2nd Sess., April 24, 1984, 180.

48. Ibid., 164.

49. William J. Broad, "Science Showmanship: A Deep Star Wars Rift," *New York Times*, December 16, 1985.

50. U.S. Senate Committee on the Armed Services, *Department of Defense Authorization for Appropriations for Fiscal Year 1985, Part 6. The Strategic Defense Initiative: Defensive Technologies Study*, 98th Cong., 2nd sess., March 8, 22, April 24, 1984, 2912.

51. Rick Atkinson, "'Star Wars" and the ASAT Projects," *Washington Post*, June 23 ,1984.

52. Brockway McMillan (chairman) et al., *Battle Management Communications, and Data Processing*, vol. 5 of *Report of the Study on Eliminating the Threat Posed by Nuclear Ballistic Missiles*, 1984), 7.

53. Ibid., 8.

54. Ibid., 8.

55. Ibid., 4.

56. House Committee on Appropriations, *Department of Defense Appropriations for 1984, Part 7*, 98th Cong., 1rst sess., May 2, 10, 18, June 8, 21, 1983, 973.

57. McMillan et al., *Battle Management, Communications, and Data Processing*, 47–48.

58. Greg Nelson and David Redell, "The Star Wars Computer System," (Palo Alto, CA: Computer Professionals for Social Responsibility, 1985).

59. Ibid.

60. House Committee on Foreign Affairs, 99th Cong., 1rst sess., *Implications of the President's Strategic Defense Initiative and Antisatellite Weapons Policy*, 99th Cong., 1rst sess., April 24, May 1, 1985, 11.Walter Mondale made similar comments in public debates: Lou Cannon, "'Star Wars' Descriptions Were Off Target; Defense Falls Far Short of the Candidates' Claims During Debate," *Washington Post*, October 23, 1984.

61. Bill Keller, "Pentagon Asserts 'Star Wars' Tests Won't Break Pact," *New York Times*, April 21, 1985.

62. See Paul Nitze, *From Hiroshima to Glasnost* (New York: Grove Weidenfeld, 1989), 412–415.

63. Charles Mohr, "U.S. Keeps Options for 'Star Wars,'" *New York Times*, October 20, 1985.

64. Charles Mohr, "U.S. Negotiators of ABM Treaty Say Reagan Is 'Harpooning" Pact," *New York Times*, October 12, 1985; Michael R. Gordon, "Nunn Says Record on ABM Pact is Being Distorted," *New York Times*, March 12, 1987.

65. Bernard Gwertzman, "Reagan Sees Hope of Soviet Sharing in Missile Defense," *New York Times*, March 30, 1983.

66. Bernard Weinraub, "How Grim Ending in Iceland Followed Hard-Won Gains," *New York Times*, October 14, 1986; George C. Wilson, "Reagan Arms Plan Questioned," *Washington Post*, October 17, 1986.

67. Jerome Wiesner and Kosta Tsipis, "Put 'Star Wars' Before a Panel," *New York Times*, November 11, 1986.

68. Ibid.

69. Kurt Gottfried and Henry W. Kendall, "Space-Based Missile Defense," (Cambridge, MA: Union of Concerned Scientists, 1984), 2.

70. Ibid.

71. Ashton Carter, "Directed Energy Missile Defense in Space: Background Paper" (Washington, DC: The Office of Technology Assessment, 1984), 81.

72. N. Bloembergen et al., "APS Study: Science and Technology of Directed Energy Weapons," *Reviews of Modern Physics* 59, no. 3 (1987): S9.

73. OTA, "Ballistic Missile Defense Technologies" (Washington, DC: U.S. Government Printing Office, 1985), 33.

74. Ibid., 128.

75. For more on Nitze's role, see Nitze, *From Hiroshima to Glasnost*, 402–403.

76. Rebecca Slayton, "Speaking As Scientists: Computer Professionals in the Star Wars Debate," *History and Technology* 19, no. 4 (2004).

77. Butterfield, "Anatomy of the Nuclear Protest."

78. R. Slayton, "Discursive Choices: Boycotting Star Wars between Science and Politics," *Social Studies of Science* 37, no. 1 (2007): 43.

79. Lin sent out a draft abstract on the Arms-D list: Herbert Lin, Software Development Effort for the Sdi—Draft Paper [ARPAnet list] (Arms Discussion Digest, 5 November 1984 [cited December 2003]); available from Posts to the Arms-Discussion Digest, beginning with Volume 5, issue 2. http://groups.google.com/group/mod .politics.arms-d/topics.

80. Barry Boehm, *Software Engineering Economics*, Prentice-Hall Advances in Computing Science and Technology Series (Englewood Cliffs, NJ: Prentice-Hall, 1981).

81. Herbert Lin, "The Development of Software for Ballistic-Missile Defense," *Scientific American* 253, no. 6 (1985): 53.

82. Parnas, "Software Aspects of Strategic Defense Systems."

83. Ibid.

84. Ibid.

85. Ibid.

86. Ibid.

87. Ibid.

88. Parnas to Slayton, e-mail, May 23, 2012.

89. Parnas to Slayton, e-mail, April 19 2003.

90. Michael Dertouzos (moderator) et al., "Star Wars: Can the Computing Require-ments Be Met?" Cambridge, MA, October 21, 1985.

91. On conflicts of interest, see Senate Committee on Governmental Affairs, *Department of Defense/Strategic Defense Initiative Organization Compliance with Federal Advisory Committee Act*, 100th Cong., 2nd sess., April 19, 1988.

92. Jim Horning, "Trip Report: Computing in Support of Battle Management," *Risks Digest*, August 21, 1985 [cited September 2003]). http://catless.ncl.ac.uk/Risks/search.html.

93. Ibid.

94. Ibid.

95. Ibid.

96. Senate Committee on Armed Services, *Strategic Defense Initiative*, 99th Cong., 1rst session, October 30; November 6, 21; December 3, 5, 1985, 53.

97. Brooks to Slayton, e-mail September 9, 2007.

98. Frederick Brooks, "No Silver Bullet: Essence and Accidents of Software Engineering," *IEEE Computer* 1987, 12.

99. Ibid., 11.

100. Senate Committee on Armed Services, *Strategic Defense Initiative*, 53–54.

101. Ibid., 286.

102. Ibid.

103. Ibid., 288.

104. Ibid., 281; Boyce Rensberger, "Computer Bugs Seen As Fatal Flaw in Star Wars," *Washington Post*, October 30, 1985.

105. Transcript of a Debate on SDI at Stanford University, Stanford, CA. December 19, 1985, 26. Files of Rodney Hoffman. Hereafter, "Stanford debate."

106. David E. Sanger, "A Debate about Star Wars," *New York Times*, October 23, 1985.

107. Stanford debate, 22.

108. Daniel Cohen, "Eastport Study Group: A Report" (SDIO, 1985), 20.

109. Ibid., 20–21.

110. Stanford debate, 17.

111. Stanford debate, 31.

112. Senate Committee on Armed Services, *Strategic Defense Initiative*, 341.

113. "Star Wars: Can the Computing Requirements Be Met?" Transcript of a debate at MIT, Cambridge, MA, 21 Oct, 1985, 47. From the files of Rodney Hoffman.

114. OTA, "SDI: Technology, Survivability, and Software" (Washington, DC: GPO, 1988), 4–5.

115. "SDIO Plans to Issue RFP for Test Facility," *Aviation Week & Space Technology*, February 17, 1985.

116. David E. Sanger, "Many Experts Doubt 'Star Wars' Could Be Effective by the Mid-90's," *New York Times*, February 11, 1987. William J. Broad, "Pentagon Starts Project to Judge Anti-Missile Plan," *New York Times*, August 16, 1987.

117. House Committee on Appropriations, *Defense Department Appropriations for 1987; Part 5*, 1986, 647.

118. Gordon, "Nunn Says Record on ABM Pact is Being Distorted"; Michael R Gordon, "Reagan Is Warned by Senator Nunn over ABM Treaty," *New York Times*, February 7, 1987.

119. "Defense Science Board Report on SDI," *Washington Post*, July 10, 1987.

120. House Armed Services Committee, *Special Panel on the Strategic Defense Initiative*, 100th Cong., 2nd sess., April 20, July 14, September 29, October 4, 1988, 277.

121. Ibid., 268–269.

122. John H. Cushman, "Pentagon Moves Toward Slowing Anti-Missile Plan," *New York Times*, May 21, 1988.

123. Senate Committee on Armed Services, *Strategic Defense Initiative*, 278–279.

124. Ibid., 280.

125. Ibid., 281.

126. Jeffrey Smith, "SDI Faulted in 2-Year Hill Study," *Washington Post*, April 24, 1988.

127. James A. Abrahamson and John H. Gibbons, "That SDI Story," *Washington Post*, May 5, 1988.

128. OTA, "SDI: Technology, Survivability, and Software," 4–5.

129. Jonathan Fuerbringer, "Budget Accords on the Military and on Tobacco," *New York Times*, December 14, 1985; Philip Boffey, "Obstacles Force a Narrower Focus on 'Star Wars'," *New York Times*, October 19, 1986.

130. Joint Committee on Armed Services, *Restructuring of the Strategic Defense Initiative (SDI) Program*, 100th Cong., 2nd sess., October 6, 1988, 2.

131. John H. Cushman, "Pentagon Official Proposes Cost Cut for Space Weapons," *New York Times*, September 8,1988.

132. William J. Broad, "What's Next for Star Wars? Brilliant Pebbles," *New York Times*, April 25, 1989.

133. For discussion of budget items, see Joint Committee on Armed Services, *Restructuring of the Strategic Defense Initiative (SDI) Program*, 32.

134. Andrew Rosenthal, "Break From Reagan," *Washington Post*, January 27, 1989.

Chapter 9

1. "Remarks by the President on Strengthening Missile Defense in Europe," September 29, 2009. http://www.whitehouse.gov/the_press_office/Remarks-by-the -President-on-Strengthening-Missile-Defense-in-Europe.

2. Arms Control Association, "Who Has What at a Glance," April 2012. http://www .armscontrol.org/factsheets/Nuclearweaponswhohaswhat#1.

3. James E. Goodby and Sidney D. Drell, "Rethinking Nuclear Deterrence." Paper presented at the Reykjavik Revisited: Steps Toward a World Free of Nuclear Weapons, Stanford, CA, 2007.

4. Frederick Brooks, "No Silver Bullet: Essence and Accidents of Software Engineering." *IEEE Computer* 20, no. 4 (1987): 12.

5. Robert Apple, "Shall We Dance?," *New York Times*, December 24, 1989.

6. Andrew Rosenthal, "Bush and Gorbachev Proclaim a New Era for U.S.-Soviet Ties," *New York Times*, December 4, 1989; Alan Riding, "Bush Says Soviets Merit West's Help to Foster Reforms," *New York Times*, December 5, 1989.

7. See, for example, R. C. McFarlane and Z. Smardz, *Special Trust* (New York: Cadell & Davies, 1994); P. Schweizer, *Victory: The Reagan Administration's Secret Strategy that Hastened the Collapse of the Soviet Union* (New York: Atlantic Monthly Press, 1996).

8. P. J. Westwick, "'Space Strike Weapons' and the Soviet Response to SDI," *Diplomatic History* 32, no. 5 (2008); Pavel Podvig, "Did Star Wars Help End the Cold War? Soviet Response to the SDI Program," unpublished manuscript (2004).

9. Jeffery Smith, "Pentagon Increases SDI Push," *Washington Post*, February 18, 1990.

10. William J. Broad, "War Hero Status Possible for the Computer Chip," *New York Times*, January 21, 1991.

11. Ibid. John Burgess, "Videotapes Show "Smart" Bombs in Action; With Sharpshooter Accuracy, Weapons Are Guided by Computers and Lasers," *Washington Post*, January 19, 1991.

12. Broad, "War Hero Status Possible for the Computer Chip."; John Lichfield, Phil Davison, and Amman Bulloch, "The West's Sophisticated Air Weaponry," *The Independent*, January 20, 1991.

13. Richard Saltus, "'Smart Bombs' Revolutionized Aerial Warfare," *Boston Globe*, January 19, 1991.

14. Peter Applebome, "The Antiwar Movement," *New York Times*, January 21, 1991. See also Susan Jeffords and Lauren Rabinovitz, eds., *Seeing Through the Media: The Persian Gulf War* (New Brunswick, NJ: Rutgers University Press, 1994).

15. Rick Atkinson and Dan Balz, "Scud Hits Tel Aviv," *Washington Post*, January 23, 1991.

16. Lawrence Freedman and Efraim Karsh, "How Kuwait Was Won: Strategy in the Gulf War," *International Security* 16, no. 2 (1991).

17. Phil Agre, "Patriot Missiles," *RISKS Digest,* January 26, 1991. http://www.catless.com/Risks/10.81.html#subj3.1.

18. Dave Parnas, "Patriot Missile," *RISKS Digest*, January 28, 1991. http://www.catless.com/Risks/10.82.html#subj1.1.

19. Clifford Johnson, "Patriots," *RISKS Digest*, January 29, 1991. http://www.catless.com/Risks/10.83.html#subj4.

20. The Patriot was only designed to operate for 24 hours at a time before rebooting, and hence the timing problem did not matter in previous operating conditions. Technically, this would be described as a "requirements failure." GAO, "Patriot Missile Defense: Software Problem Led to System Failure at Dhahan, Saudi Arabia" (Washington, DC: General Accounting Office, 1992).

21. Juan J. Walte, "Desert Storm Was Test Lab for Patriots," *USA Today*, May 24, 1991.

22. GAO, "Patriot Missile Defense: Software Problem Led to System Failure at Dhahan, Saudi Arabia."

23. For more on Lewis' work, see R. Slayton, "Discursive Choices: Boycotting Star Wars Between Science and Politics," *Social Studies of Science* 37, no. 1 (2007).

24. George N. Lewis and Theodore A. Postol, "Video Evidence on the Effectiveness of Patriot during the 1991 Gulf War," *Science & Global Security* 4 (1993).

25. See George N. Lewis and Theodore A. Postol, "Technical Debate over Patriot Performance in the Gulf War," *Science & Global Security* 3 (2000).

26. James Asker, "Ballistic Missile Defense Shifts to Cover Theater/Tactical Threats," *Aviation Week*, March 18, 1991.

27. Jeffery Smith and Dan Morgan, "SDI Ordered to Aim Lower, Scale Back," *Washington Post*, January 31, 1991.

28. Ibid. Asker, "Ballistic Missile Defense Shifts To Cover Theater/Tactical Threats."

29. Senate Committee on Foreign Relations, *The SDI As It Relates to the ABM Treaty*, 102nd Cong., 1rst sess., April 24, 1991.

30. Quoted in Graham Spinardi, "Ballistic Missile Defence and the Performativity Problem: The Development of Hit-to-kill Technology," unpublished manuscript (2010).

31. Ibid.

32. Richard Garwin, "Defense Is Easier from the Ground," *Space News*, March 11–17, 1991; "Brilliant Pebbles Won't Work Against Theater Missiles, Group Says," *Aerospace Daily*, March 14, 1991.

33. Don Oberdorfer, "U.S., Russia to Deepen A-Arms Cut," *Washington Post*, June 9, 1992; Don Oberdorfer and Jeffery Smith, "New Era of Nuclear Disarmament," *Washington Post*, February 2, 1992; Fred Hiatt, "Russian Legislature Ratifies START Pact," *Washington Post*, November 5, 1992.

34. Erik K. Pratt, "Missile Defense Sponsors: Shifting Political Support for Strategic Defense after Reagan," *Asian Perspective* 25, no. 1 (2001).

35. Quoted in ibid., 22.

36. Donald Baucom, "The Rise and Fall of Brilliant Pebbles," *Journal of Social, Political and Economic Studies* 29, no. 2 (2004).

37. James Asker, "SDI Budget Shifts to 'Here and Now'," *Aviation Week & Space Technology*, April 12, 1993; "SDIO, in shift to theater defenses, slips NMD deployment," *Aerospace Daily*, March 30, 1993; "SDI's Priorities," *Aviation Week & Space Technology*, March 1, 1993.

38. David Hughes, "BMDO Under Pressure to Set TMD Priorities," *Aviation Week & Space Technology*, January 17 1994.

39. David Fulgham, "U.S. Missile Defense Plans Narrowed," *Aviation Week & Space Technology*, December 4, 1995.

40. "BMDO Sees Slow, Steady Budget Climb Through Fiscal '99," *Aerospace Daily*, March16 1994; Editor, "Pentagon Delays Land-Based Interceptor," *Aviation Week & Space Technology*, March 1, 1993.

41. GAO, "Review of Allegations about an Early National Missile Defense Flight Test" (Washington, DC: GAO, 2002), 6.

42. Gary Taubes, "Postol vs. the Pentagon," *Technology Review*, April 2002, 57.

43. William Broad, "Missile Contractor Doctored Tests, Ex-Employee Charges," *New York Times*, March 7, 2000.

44. Taubes, "Postol vs. the Pentagon." Ted Postol and George Lewis, "Future Challenges to Ballistic Missile Defense," *IEEE Spectrum* (1997): 68.

45. GAO, "Review of Allegations about an Early National Missile Defense Flight Test," 6.

46. GAO, "Missile Defense: Events Related to Contractor Selection for the Exoatmospheric Kill Vehicle—Report to Howard L. Berman, House of Representatives, from Anthony Gamboa, GAO General Counsel," (Washington, DC: GAO, 2003).

47. GAO, "DOD's Goals for Resolving Space Based Infrared System Software Problems Are Ambitious" (Washington, DC: GAO, 2003); GAO, "Space-Based Infrared System-Low at Risk of Missing Initial Deployment Date" (Washington, DC: GAO, 2001).

48. GAO, "Missile Defense: Events Related to Contractor Selection for the Exoatmospheric Kill."

49. Pratt, "Missile Defense Sponsors: Shifting Political Support for Strategic Defense after Reagan."

50. Michael Dobbs, "How Politics Helped Redefine Threat," *Washington Post*, January 14, 2002.

51. Ibid.

52. Ibid.

53. Pratt, "Missile Defense Sponsors: Shifting Political Support for Strategic Defense after Reagan."

54. GAO, "National Missile Defense: Even with Increased Funding Technical and Schedule Risks Are High," (Washington, DC: General Accounting Office, 1998), 6.

55. Ibid.

56. William Scott, "Mix of Simulation, Flight Testing Troubles BMDO Leaders," *Aviation Week & Space Technology*, February 24, 1997.

57. Senate Armed Services Committee, *Department of Defense Authorization for Appropriations for Fiscal Year 1999 and the Future Years Defense Program*, 105th Cong., 2nd Sess., March 11, 12, 19, 24, 26, 31, 1998, 390.

58. Scott, "Mix of Simulation, Flight Testing Troubles BMDO Leaders."

59. William Scott, "National Test Facility Speeds Development of Simulation for Missile Defense System," *Aviation Week & Space Technology*, February 17, 1992; William Scott, "NTF Broadens Scope of Test Activities," *Aviation Week & Space Technology*, May 9, 1994; Scott, "Mix of Simulation, Flight Testing Troubles BMDO Leaders."

60. William Scott, "Reality Checks Boost Wargame Credibility," *Aviation Week & Space Technology*, November 2, 1998.

61. House Committee on Government Reform, *National Missile Defense: Test Failures and Technology Development*, 106th Cong., 2nd Sess., September 8, 2000, 69.

62. Ibid., 72.

63. Ibid., 87.

64. *Department of Defense Authorization for Appropriations for Fiscal Year 1999 and the Future Years Defense Program*, 390.

65. Ibid.

66. Bradley Graham, "Panel Faults Antimissile Program on Many Fronts," *Washington Post*, November 14, 1999.

67. Dana Priest, "Cohen Says U.S. Will Build Missile Defense," *Washington Post*, January 21, 1999.

68. Michael Dornheim, "National Missile Defense Focused on June Review," *Aviation Week & Space Technology*, August 16, 1999.

69. Andrew M. Sessler (chair of the Study Group) et al., "Countermeasures: A Technical Evaluation of the Operational Effectiveness of the Planned US National Missile Defense System" (Cambridge, MA: Union of Concerned Scientists, MIT Security Studies Program, 2000), 105–106.

70. See Elaine Sciolino, "Antimissile Test Fails Over Pacific," *New York Times*, July 8, 2000.

71. David Hoffman, "Russia Says START II Is Imperilled," *Washington Post*, January 22, 1999.

72. Pratt, "Missile Defense Sponsors: Shifting Political Support for Strategic Defense after Reagan."

73. David Hoffman, "Moscow Proposes Extensive Arms Cuts," *Washington Post*, August 20, 1999.

74. David Hoffman, "Russia Apparently Snubs U.S. Radar Offer," *Washington Post*, October 20, 1999.

75. Charles Babington, "U.S. Set to Share Its ABM Research," *Washington Post*, June 1, 2000.

76. Sharon LaFraniere, "Missile Shield under Attack," *Washington Post*, July 6, 2000.

77. William J. Broad, "Clinton's Missile Decision," *New York Times*, September 2, 2000.

78. House Committee on Government Reform, *National Missile Defense: Test Failures and Technology Development*.

79. Michael R. Gordon and Steven Lee Meyers, "Bush Team Vows to Speed up Work on Missile Shield," *New York Times*, April 30, 2001.

80. David E. Sanger and Steven Lee Meyers, "Bush's Missile Plan," *New York Times*, May 2, 2001.

81. Frank Bruni, "Putin Urges Bush Not to Act Alone on Missile Shield," *New York Times*, June 17, 2001; Frank Bruni, "France and Germany Caution Bush on Missile Defense Plan " *New York Times*, June 14, 2001.

82. Quoted in David E. Sanger and Patrick Tyler, "Officials Recount Road to Deadlock over Missile Talks," *New York Times*, December 13, 2001.

83. Quoted in ibid.

84. Steven Mufson and Sharon LaFraniere, "ABM Withdrawal: A Turning Point in Arms Control," *Washington Post*, December 13, 2001.

85. Michael Wines, "After U.S. Scraps ABM Treaty, Russia Rejects Curbs of START II," *New York Times*, June 15, 2002.

86. Helen Dewar, "Missile Defense Funding Increased," *Washington Post*, June 27, 2002.

87. Bradley Graham, "Missile Defense to Start in 2004," *Washington Post*, December 18, 2002.

88. "Anger as Hoon Gives Backing to Son of Star Wars," *The Herald (Glasgow)*, January 16, 2003; Jeff Sallot, "U.S. Planning Space Weapons, Russian Envoy Says," *The Globe and Mail* (Canada), August 20, 2004.

89. Michael Gordon, "U.S. Is Proposing European Shield For Iran Missiles," *New York Times*, May 22, 2006; Thom Shanker, "Russian Criticizes U.S. Plan For Missile Defense System," *New York Times*, February 10, 2007; Thom Shanker, "Poland Ties U.S. Missile Plan to Security Pledges," *New York Times*, April 25, 2007; ibid.

90. Steven Erlanger and Steven Lee Myers, "NATO Backs Missile Defense in Europe, but Rejects Admitting Georgia and Ukraine," *New York Times*, April 4, 2008.

91. Michael Schwirtz, Anne Barnard, and C. J. Chivers, "Russia and Georgia Clash Over Breakaway Region," *New York Times*, August 9, 2008.

92. Andrew Kramer and Ellen Barry, "Russia, in Accord with Georgians, Sets Withdrawal," *New York Times*, August 13, 2008.

93. Clifford Levy, "Despite Pullout, Russia Envisions Long-Term Shift," *New York Times*, August 23, 2008.

94. Clifford Levy, "Russia Vows to Support Two Enclaves, in Retort to Bush," *New York Times*, August 23, 2008; Steven Lee Myers, "Bush, Sending Aid, Demands Moscow Withdraw Forces," *New York Times*, August 14, 2008.

95. Karen DeYoung, "U.S. and Poland Seal Missile Pact," *Washington Post*, August 21, 2008; Thom Shanker and Nicholas Kulish, "U.S. and Poland Set Missile Deal," *New York Times*, August 15, 2008.

96. Quoted in Lisbeth Gronlund et al., "Technical Realities: An Analysis of the 2004 Deployment of a U.S. National Missile Defense System" (Cambridge, MA: Union of Concerned Scientists, MIT Security Studies Program, 2004), 25.

97. Bradley Graham, "Missile Defense Choices Sought," *Washington Post*, September 3, 2002; David E Sanger, "Bush Issues Directive Describing Policy on Antimissile Defenses," *New York Times*, May 21, 2003.

98. For more on the history of defense acquisition policies, and discussion of evolutionary development processes, see http://history.defense.gov/acqh.shtml.

99. Graham, "Missile Defense Choices Sought."

100. Graham, "Missile Defense to Start in 2004."

101. Sessler et al., "Countermeasures: A Technical Evaluation of the Operational Effectiveness of the Planned US National Missile Defense System," 1–2.

102. Gronlund et al., "Technical Realities: An Analysis of the 2004 Deployment of a U.S. National Missile Defense System," ix.

103. David Stout and John Cushman, "Defense Missile for U.S. System Fails to Launch," *New York Times*, December 16, 2004.

104. Surveying the history of testing EKVs from the Army's HOE through the 2000 ground-based midcourse program, twenty tests of space-based interceptors scored five successful intercepts. See Sessler et al., "Countermeasures: A Technical Evaluation of the Operational Effectiveness of the Planned US National Missile Defense System," 1–2.

105. Bradley Graham, "'Minor' Software Glitch Is Cited in Missile Failure," *Washington Post*, January 13, 2005.

106. Ibid.

107. GAO, "Missile Defense Agency Fields Initial Capability but Falls Short of Original Goals" (Washington, DC: GAO, 2006), 3.

108. GAO, "Assessment of Progress Made on Block 2006 Missile Defense Capabilities and Oversight" (Washington, DC: GAO, 2008), 2.

109. Ibid.

110. GAO, "Ballistic Missile Defense: Actions Needed to Improve Process for Identifying and Addressing Combatant Command Priorities" (Washington, DC. GAO, 2008).

111. Ellen Barry and Sophia Kishkovsky, "Russian President Sends Obama Warning on European Missile System," *New York Times*, November 6, 2008.

112. Peter Baker, "Obama Reshapes a Missile Shield to Blunt Tehran," *New York Times*, September 18, 2009.

113. Peter Baker, "With Arms Treaty, A Challenge Remains," *New York Times*, April 8, 2010.

114. Ellen Barry, "Senate Rejects Amendment Blocking New Start Treaty," *New York Times*, December 19, 2010.

115. Walter Pincus, "Cold War Issues Linger over Arms-Control Negotiations," *Washington Post*, January 18, 2011.

116. Fred Weir, "New U.S.-Russia Arms Race? Battle Lines Grow over Missile Defense," *Christian Science Monitor*, June 8, 2011.

117. Ted Postol and Yousaf Butt, "Upsetting the Reset: The Technical Basis of Russian Concern Over NATO Missile Defense" (FAS Special Report No. 1, 2011).

118. House Committee on Armed Services, *Report on the Ballistic Missile Defense Review and the Fiscal Year 2011 National Defense Authorization Budget Request for Missile Defense Programs*, 111th Cong., 2nd Sess., April 15, 2010, 47. See also March 3, 2010, fact sheet, http://www.mda.mil/global/documents/pdf/bmdr_fs.pdf.

119. http://www.mda.mil/global/documents/pdf/testrecord.pdf.

120. See Amy Butler, "Sbirs Satellite Activated," *Aerospace Daily & Defense Report*, July 11, 2011. See also GAO, "DOD Delivering New Generations of Satellites but Space System Acquisition Challenges Remain" (Washington, DC: GAO, 2011).

121. White House Office of the Press Secretary, "Fact Sheet on U.S. Missile Defense Policy," September 17, 2009. http://www.whitehouse.gov/the_press_office/FACT -SHEET-US-Missile-Defense-Policy-A-Phased-Adaptive-Approach-for-Missile -Defense-in-Europe,

122. Ted Postol and George Lewis, "How US Strategic Antimissile Defense Could Be Made to Work," *Bulletin of Atomic Scientists* 66, no. 6 (2010).

123. Welch, Larry, and Robert Hermann (co-chairs), "Defense Science Board Task Force on Science and Technology Issues of Early Intercept Ballistic Missile Defense Feasibility" (U.S. Department of Defense, 2011), 27.

124. Robert Wall and Douglas Barrie, "Extending the Shield," *Aviation Week & Space Technology*, October 3, 2005. The U.S. is currently trying to end its involvement in MEADS, to the consternation of allies. See, for example, Kate Brannon, "Italian Defense Minister to U.S.: Respect MEADS Commitment," DefenseNews.com, April 30, 2012.

125. Missile Defense Agency, "Terminal High Altitude Area Defense," http://www .mda.mil/system/thaad.html. THAAD currently is only deployed in two batteries at Ft Bliss, Texas.

126. Michael A. Taverna, "Out of Denial," *Aviation Week & Space Technology* 172, no. 4 (2010).

127. NATO Provides Information about the Evolution of Its Active Layered Theater Ballistic Missile Defense (ALTBMD): http://www.nato.int/cps/en/natolive/topics_49635.htm.

128. GAO, "Missile Defense: European Phased Adaptive Approach Acquisitions Face Synchronization, Transparency, and Accountability Challenges" (Washington, DC: GAO, 2010), 3.

129. N. G. Leveson, "Software Safety: Why, What, and How," *Computing Surveys* 18, no. 2 (1986): 135–136.

130. Ibid.

131. See, for example, Committee for Advancing Software-Intensive Systems Producibility, National Research Council, *Critical Code: Software Producibility for Defense* (Washington, DC:, 2010), 54. Daniel Jackson, Martyn Thomas, and Lynette Millett, eds., *Software for Dependable Systems* (Washington, DC: National Academies Press, 2007).

132. The IEEE Computer Society provides ongoing news about certification; see http://www.computer.org/portal/web/certification/home.

133. J. C. Knight and N. G. Leveson, "Should Software Engineers Be Licensed?," *Communications of the ACM* 45, no. 11 (2002): 89.

134. Ibid.; John White and Barbara Simons, "ACM's Position on the Licensing of Software Engineers" *Communications of the ACM*, 45, no 11 (2002): 91.

135. For an example of how engineers try to manage this process, see T. P. Kelly, "A Systematic Approach to Safety Case Management" (SAE Technical Paper 2004–01–1779, 2004).

136. Frederick Brooks et al., "Panel: 'No Silver Bullet' Reloaded." Paper presented at the 22nd Annual ACM SIGPLAN Conference on Object-Oriented Programming, Systems, Languages, and Applications (OOPSLA), Montreal, Canada, 2007, 1029.

137. Fred Brooks (chairman), "Report of the Defense Science Board Task Force on Military Software" (U.S. Department of Defense, 1987), 16.

138. Ibid., 1.

139. Craig Fields, "Report of the Defense Science Board Task Force on Defense Software" (Washington, DC: Defense Department, 2000), ES1.

140. See, for example, George H. Heilmeier and Larry Druffel, "Report of the Defense Science Board Task Force on Acquiring Defense Software Commercially" (Washington, DC: Defense Department, 1994).

141. Brooks, "Report of the Defense Science Board Task Force on Military Software."

142. Mark Hillman, "Just One Stock: Solid Core, New Products Set Up Refreshing Growth for Microsoft, May 10, 2011. http://seekingalpha.com/article/269165-just -one-stock-solid-core-new-products-set-up-refreshing-growth-for-microsoft.

143. Gregg Keizer, "Apple Breaks iPhone Sales Record Again," Computerworld .com, April 20 2011. http://www.computerworld.com/s/article/9216004/Apple _breaks_iPhone_sales_record_again.

144. Fields, "Report of the Defense Science Board Task Force on Defense Software."

145. Committee for Advancing Software-Intensive Systems Producibility, National Research Council, *Critical Code: Software Producibility for Defense*," 19.

146. Ibid., 2.

147. Brooks, "Report of the Defense Science Board Task Force on Military Software," 33. Committee on Improving Processes and Policies for the Acquisition and Test of Information Technologies in the Department of Defense, National Research Council, *Achieving Effective Acquisition of Information Technology in the Department of Defense* (Washington, DC: National Academies Press, 2010), 50. Committee for Advancing Software-Intensive Systems Producibility, *Critical Code: Software Producibility for Defense*," 60.

148. In 2009, an Obama-Biden team on acquisition reform noted that the current system is "fundamentally broken." Quoted in Committee on Improving Processes and Policies for the Acquisition and Test of Information Technologies in the Department of Defense, *Achieving Effective Acquisition of Information Technology in the Department of Defense*," 29.

149. Edward Adams, "Optimizing Preventive Service of Software Products," *IBM Journal of Research and Development* 28, no. 1 (1984).

150. William J. Broad and Carl Hulse, "The Great-Grandson of Star Wars, Now Ground-Based, is Back on the Agenda," *New York Times*, June 8, 2004.

151. See, for example, "China Confirms Satellite Downed," BBCnews.com, January 23, 2007. http://news.bbc.co.uk/2/hi/asia-pacific/6289519.stm.

152. According to a 2000 estimate, about 5 percent of satellites fail in their first 6 to 12 months in orbit. Allen Gould and Orin Linden, "Estimating Satellite Insurance Liabilities," *Casualty Actuarial Society Forum*, Fall 2000, 44–87. http://www.casact .org/pubs/forum/00fforum/00ff047.pdf.

153. Kadish testimony, House Committee on Appropriations, *Department of Defense Appropriations*, 108th Cong., 1rst Sess., March 13, 27, April 30, May 1, September 30, 2003.

154. Jonathan Weisman, "Patriot Missiles Seemingly Falter for Second Time; Glitch in Software Suspected," *Washington Post*, March 26, 2003.

155. Bradley Graham, "Radar Probed in Patriot Incidents," *Washington Post*, May 8, 2003.

156. Michael Williams and William Delaney, "Report of the Defense Science Board Task Force on Patriot System Performance," (Washington, DC: Office of the Under Secretary of Defense for Acquisition, Technology, and Logistics, 2005).

157. GAO, "Missile Defense: European Phased Adaptive Approach Acquisitions Face Synchronization, Transparency, and Accountability Challenges."

158. Williams and Delaney, "Report of the Defense Science Board Task Force on Patriot System Performance."

159. Thom Shanker and Eric Schmitt, "Rumsfeld Orders War Plans Redone for Faster Action," *New York Times*, October 13, 2002.

160. The transcript is available from ABCnews at http://abcnews.go.com/ International/story?id=79503&page=1.

161. See, for example, Carl Robinchaud, "Failings of the Rumsfeld Doctrine," *Christian Science Monitor*, September 21, 2006. http://www.csmonitor.com/2006/0921/ p09s02-coop.html.

162. House Committee on Government Reform, *National Missile Defense: Test Failures and Technology Development*, 105.

163. Ibid.

164. *Department of Defense Appropriations*, 180.

165. Ibid.

166. Thomas Ricks, "Investigation Finds U.S. Missiles Downed Navy Jet," *Washington Post*, December 11, 2003. UK Ministry of Defense, "Aircraft Accident to Royal Air Force Tornado GR MK4A ZG710," May 2004.

167. http://www.raytheon.com/capabilities/products/patriot/pat_home.

168. W. Yurcik and D. Doss, "Software Technology Issues for a US National Missile Defense System," *IEEE Technology and Society Magazine* 21, no. 2 (2002).

169. Mark Halpern, "Buggy Software and Missile Defense," *New Atlantis*, no. 10 (2005).

170. See discussion in chapter 7 and David Parnas, "Designing Software for Ease of Extension and Contraction," *IEEE Transactions on Software Engineering* (1979); David Parnas, "On the Criteria to Be Used in Decomposing Systems into Modules," *Communications of the ACM* 15, no. 12 (1972).

171. Missile Defense Agency, "Historical Funding for MDA FY85-12," http://www.mda.mil/global/documents/pdf/histfunds.pdf.

172. On poverty, see Sabrina Tavernise, "Soaring Poverty Casts Spotlight on 'Lost Decade.'" http://www.nytimes.com/2011/09/14/us/14census.html?_r=3&hpw&. In 2011, the MDA was allocated $8.5 billion; in 2012 it requested $8.6 billion. See http://www.mda.mil/global/documents/pdf/budgetfy12.pdf and http://www.mda.mil/global/documents/pdf/histfunds.pdf.

173. Charles Perrow, *The Next Catastrophe: Reducing Our Vulnerabilities to Natural, Industrial, and Terrorist Disasters* (Princeton, NJ: Princeton University Press, 2007).

174. D. K. Barton et al., "Report of the American Physical Society Study Group on Boost-Phase Intercept Systems for National Missile Defense: Scientific and Technical Issues," *Reviews of Modern Physics* 76, no. 3 (2004): xxi–xxii. See also Sessler et al., "Countermeasures: A Technical Evaluation of the Operational Effectiveness of the Planned US National Missile Defense System," 1–2.

175. Postol and Lewis, "Future Challenges to Ballistic Missile Defense," 68.

176. Dan Auerbach and Lee Tien, "Dangerously Vague Cybersecurity Legislation Threatens Civil Liberties," March 30, 2012. https://www.eff.org/deeplinks/2012/03/dangerously-vague-cybersecurity-legislation.

177. "Hackers in Demand at US Government Agencies," *The Telegraph*, August 3, 2011. http://www.telegraph.co.uk/technology/news/8679293/Hackers-in-demand-at-US-government-agencies.html.

178. Licklider to Wiesner, "Underestimates and Overexpectations in the Development of Complex Systems," 24 March 1969, MC420, Box 19, Folder 11. See discussion in chapter 5.

179. See, for example, National Research Council, *Biotechnology Research in an Age of Terrorism* (National Academies Press, 2004).

Unpublished Sources and Notations

ARCHIVES

Neils Bohr Library, American Institute of Physics, College Park, MD.
• Records of the American Physical Society, Directed Energy Weapons Study, 1983–1988: *APS DEW Records*.
• Records of Irwin Goodwin, 1983–1993. *Irwin Goodwin Papers*.

Division of Rare and Manuscript Collections, Carl A. Kroch Library, Cornell University, Ithaca, NY.
• Hans Bethe papers, ca. 1931–1992, Collection Number 14–22–976: *Bethe Papers*.

Hoover Institution Archives, Stanford University, Stanford, CA.
• Edward Teller Papers, 1954–1997: *Teller Papers*.

Mandeville Special Collections Library, UCSD Libraries, LaJolla, CA.
• Herbert F. York Papers: *York Papers*.

Institute Archives and Special Collections, MIT Libraries, Cambridge, MA.
• Office of the President (Compton-Killian); Records, 1930–1959 (AC4): *AC4, MIT*.
• Office of the Chancellor (Stratton); Records, 1949–1957 (AC 132): *AC132, MIT*.
• Office of the President (Stratton); Records, 1957–1966 (AC 134): *AC134, MIT*.
• Jerome B. Wiesner; Papers, 1949–1983 (MC 420): *MC420, MIT*.
• J. C. R. Licklider; Papers, 1938–1995 (MC 499): *MC499, MIT*.
• Union of Concerned Scientists (MC 434): *MC434, MIT*.
• Project Whirlwind Collection (MC 665): *MC665, MIT*.

MIT Dome, Project Whirlwind Reports, online and searchable at http://dome.mit.edu/handle/1721.3/37456: *PW Dome*.

Charles Babbage Institute, University of Minnesota, Minneapolis, Oral History Collection online and searchable at http://www.cbi.umn.edu/oh/: *CBI.*
Library of Congress, Manuscript Division, Washington, DC.

- Papers of Vannevar Bush: *Bush Papers, LC.*
- Papers of I. I. Rabi 1899–1989 (bulk 1945–1968): *Rabi Papers, LC.*
- Papers of Louis N. Ridenour: *Ridenour Papers, LC.*

National Archives and Records Administration, Washington, DC.

- Records of the Office of Science and Technology Policy, Record Group 359: *RG359, NARA.*

The National Security Archive, George Washington University, Washington, DC.

- Nuclear History Collection (Fred Kaplan Donation): *Kaplan Collection.*
- Nuclear History Program Collections, Berlin Crisis: *NHP/Berlin Crisis.*

Digital National Security Archive, online and searchable at http://nsarchive
.chadwyck.com/cat/search.do?clear=true.

- Kissinger Transcripts: *DNSA KTxxxxx (document number).*
- Military Uses of Space: *DNSA MSxxxxx (document number).*
- Nuclear History Collection: *DNSA NHxxxxx (document number).*
- Presidential Directives: *DNSA PDxxxxx (document number).*

PERSONAL FILES

Lisbeth Gronlund and David Wright, five boxes loaned to author and returned to owners in 2002–04: *Gronlund/Wright Files.*
Daniel D. McCracken, five boxes loaned to author. Currently in the Charles Babbage Institute: *McCracken Files.*

INTERVIEWS CONDUCTED

Barry W. Boehm, University of Southern California, Los Angeles, July 3, 2007.
Sidney Drell, Stanford, CA, August 2002 and August 9, 2006.
Richard Garwin, La Jolla, CA, June 28, 2007.
Marvin Goldberger, La Jolla, CA, June 28, 2007.
Kurt Gottfried, phone interview, July 2003.
Lisbeth Gronlund and David Wright, Cambridge, MA, May 26, 2004.
Spurgeon Keeney, Washington, DC, November 20, 2007.
John Kogut, University of Illinois, Urbana-Champaign, July 20, 2004.

Herbert Lin, Boston, MA, May 22, 2003.
John McCarthy, Stanford, CA, August 2002.
Brockway McMillan, Sedgwick, ME, September 2, 2006.
Peter Neumann, phone interview, September 25, 2007.
Wolfgang Panofsky, Stanford, CA, July 12, 2006.
George Rathjens, Cambridge, MA, September 1, 2006.
Jack Ruina, Cambridge, MA, December 9, 2003 and October 8, 2007.
Charles Seitz, Pasadena, CA, August 2002.
Albert (Bud) Wheelon, Montecito, CA, October 1, 2006.
Herbert York, LaJolla, CA, May 2, 2005; August 4, 2006.

Index

Printed in the United States
by Baker & Taylor Publisher Services